Relaxations and Solutions

for the

Minimum Graph Bisection Problem

Zur Erlangung des akademischen Grades eines
Doktor der Naturwissenschaften
(Dr. rer. nat.)
vom Fachbereich Mathematik
der Technischen Universität Darmstadt
genehmigte

Dissertation

von

Dipl.-Math. Marzena Fügenschuh

aus Lublin/Polen

Referent: Prof. Dr. Alexander Martin
Korreferent: Prof. Dr. Christoph Helmberg

Tag der Einreichung: 12. Juli 2007
Tag der mündlichen Prüfung: 14. September 2007

Darmstadt, 2007
D 17

Bibliografische Information der Deutschen Nationalbibliothek

Die Deutsche Nationalbibliothek verzeichnet diese Publikation in der
Deutschen Nationalbibliografie; detaillierte bibliografische Daten sind
im Internet über http://dnb.d-nb.de abrufbar.

ISBN 978-3-8325-1735-9

Logos Verlag Berlin GmbH
Comeniushof, Gubener Str. 47,
10243 Berlin
Tel.: +49 030 42 85 10 90
Fax: +49 030 42 85 10 92
INTERNET: http://www.logos-verlag.de

Jednej rzeczy mnie matematyka nauczyła
- pokory.

(One thing mathematics taught me
- the unassumingness.)

From Z. to A.

Zusammenfassung der Dissertationsschrift

Graphenpartitionsprobleme sind ein klassisches Forschungsgebiet der diskreten Optimierung. Sie finden Anwendung in der Numerik, im VLSI-Design, im Compiler-Design oder in der Frequenzzuweisung, um nur einige Bereiche zu nennen. Im Allgemeinen sind die Probleme NP-schwer. Es besteht daher wenig Hoffnung, Algorithmen mit polynomialer Laufzeit zu finden. Die große Zahl der Anwendungen führte zur Entwicklung entsprechend vieler Heuristiken. Diese Verfahren dienen der schnellen Erzeugung brauchbarer Lösungen, garantieren in der Regel jedoch nicht, eine beweisbar optimale Lösung zu finden. Gefragt sind daher gute Schranken für den Optimalwert einer Lösung. Der dominierende Ansatz zur Berechnung solcher Schranken ist die Betrachtung der konvexen, insbesondere linearen und neuerdings auch semidefiniten Relaxierung, sowie die Untersuchung des zugrunde liegenden Polyeders. Daraus abgeleitete Erkenntnisse fließen in Branch-and-Cut Verfahren in Form zusätzlicher Schnittebenen ein. In der Dissertation wird das Branch-And-Cut Verfahren basierend auf der einerseits linearen und anderseits der semidefiniten Relaxierung anhand eines Graphenpartitionierungsproblems verglichen.

Wir betrachten das *Minimum-Graphenbisektionsproblem*. Gegeben seien ein Graph mit positiven Knoten- und Kantengewichten und eine Schranke B, die mindestens die Hälfte des Knotengesamtgewichts beträgt. Gesucht wird eine Partitionierung der Knotenmenge in zwei Teilmengen, so dass das Knotengesamtgewicht jeder Teilmenge höchstens B erreicht, und die gewichtete Summe derjenigen Kanten, die Knoten in unterschiedlichen Teilmengen verbinden, minimal ist.

Unsere Arbeit basiert auf polyedrischen Studien bezüglich der Probleme Max-Cut, Equicut, Cliquenpartitionierung und Partitionierung in mehrere kapazitätsbeschränkte Knotenteilmengen. Wir verstärken die bekannten gültigen Ungleichungen unter Ausnutzung der Bisektionsbedingung zur Bestimmung von Schranken für das Minimum-Graphenbisektionsproblem. Insbesondere betrachten wir die Kreis- und Baumungleichungen, die zu *komplementären Kreis- und Baumungleichungen* führen. Der wichtigste Beitrag der Arbeit betrifft die Rucksack-Baum-Ungleichungen, eingeführt von Ferreira u.a. in 1996. Diese Ungleichungen bauen auf der Rucksackbedingung bezüglich der Knotenteilmengen auf. Empirisch erwiesen sie sich als sehr effizient. Allerdings gab es bis jetzt keine theoretische Begründung für dieses Verhalten. Wir geben notwendige und hinreichende Bedingungen an, wann die Ungleichungen Facetten des Polytops definieren.

Im praktischen Teil der Arbeit vergleichen wir, wie die neu gewonnenen polyedrischen Erkenntnisse die lineare und semidefinite Relaxierung des Minimum-Graphenbisektionproblems beeinflussen. Wir entwickeln Separierungsalgorithmen für die neuen Ungleichungen und integrieren sie in einen mit Lösern für lineare und semidefinite Programme ausgestatteten Brach-and-Cut Algorithmus. Weiterhin setzten wir heuristische Verfahren ein, die die Lösung der Relaxierungen ausnutzen. Hier wenden wir Zufallsmethoden wie genetische Algorithmen oder GRASP (Greedy Randomized Adaptive Search Procedure) an. Als Testinstanzen dienen Graphen, die aus dem VLSI-Design und Nummerik stammen, sowie zufällig erstellte Graphen.

Diese Arbeit wurde von der Deutschen Forschungsgemeinschaft unterstützt (Projektnummer MA 1324/2).

Preface

Graph partitioning problems belong to a classical research topic in discrete optimization. They have a rich practical background in scientific computing, VLSI-design, compiler design or frequency assignment, to mention just some examples. Generally these problems are intractable: so far no algorithm is known which could solve them in polynomial time. Nevertheless, a big number of applications forced the development of many heuristic methods which quickly generate a feasible solution but do not guarantee its quality in the sense of optimality. To this end, good bounds on the optimal value of a solution are in demand. A common approach to find such bounds is to consider the convex relaxation, especially linear and recently also semidefinite, and to investigate the underlying polyhedra. The outcome of this approach are cutting planes, which are integrated in a branch-and-cut framework. We apply this method to a specialized graph partitioning problem and test the theoretical results on a branch-and-cut method based on both linear and semidefinite relaxations.

In this thesis we consider the *minimum graph bisection problem*. We are given a graph with positive edge and node weights and some bound B which amounts to at least the half of the total node weight of the graph. We search for a partition of the node set into two subsets, so that the node weight of each subset stays below the bound B and the weighted sum of edges joining nodes in different clusters is minimal.

This work bases on polyhedral studies concerning the problems max-cut, equicut and partitioning into several capacity limited node subsets. We strengthen known valid inequalities with respect to the bisection condition to obtain bounds for the minimum graph bisection problem. Especially we consider tree and cycle inequalities which we upgrade to *complementarity tree and cycle inequalities*. The main contribution of this work concerns further investigations on the knapsack tree inequalities, introduced by Ferreira et al. in 1996. These inequalities base on the knapsack condition with respect to the node capacity of the subsets. Ferreira et al. empirically showed the efficiency of these inequalities. However, the theoretical investigation was missing so far. We state sufficient and necessary conditions for this class of inequalities to induce facets of the corresponding polytope, and so justify this behavior.

In the empirical part of the thesis we compare the impact of the new polyhedral results on the linear and semidefinite relaxation of the minimum graph bisection problem. We develop separation algorithms for the new inequalities and integrate them in a branch-and-cut framework featured with linear and semidefinite relaxation. Moreover, we implement heuristic methods which exploit solutions of the relaxations to construct upper bounds. Here we pursue randomized methods like genetic algorithms or greedy randomized adaptive search procedures. The graph instances we use in the computations originate in VLSI-design, scientific simulations, or are generated randomly.

This work was supported by the German Research Foundation (DFG, MA 1324/2).

Acknowledgments

First of all, I wish to thank Prof. Dr. Alexander Martin for giving me the graduate position at Darmstadt University of Technology. He bravely entrusted this research project to me and supported me with his advice, experience, and patience.

The most important chapter of this thesis would surely lose out on its readability without invaluable advice of Michael Armbruster, my project partner from Chemnitz University of Technology. He read the first draft of my proofs and sent back a list of suggestions and remarks longer than my draft at that time. I admire his efforts and endurance in incorporating the SDP-solver in SCIP and am very pleased with our cooperation over the last three years.

Many thanks to Prof. Dr. Christoph Helmberg from Chemnitz University of Technology for fruitful discussions during our meetings in Darmstadt and Chemnitz.

I am deeply grateful to Tobias Achterberg from Zuse Institute Berlin for his great support and advice on SCIP. Being himself busy with his PhD thesis he always found the time to answer quickly and in great detail surely sometimes annoying questions.

A special thank to my dear friend Dr. Heidrun Pühl, who patiently read the entire manuscript and provided me with her valuable advice.

I also would like to thank Margit Matt and her colleagues from the mathematical computing center at the Chemnitz University of Technology, who kindly allowed me to carry out my computations on their computers and provided me with necessary help.

Also many thanks to all my colleagues from the optimization group and the third floor of our "Mathebau". It was a pleasure to work with you. I will miss the coffee breaks after lunch.

Last but not least I would like to thank my husband Armin, who encouraged me to turn back from industry to research. He convinced me that applied math is enjoyable by giving me the introduction into discrete optimization on our hiking tours in Odenwald. His love and belief in me assisted me throughout the dissertation time.
Thankfully I dedicate this work to him.

Darmstadt, July 2007   *Marzena Fügenschuh*

Contents

Chapter 1

Introduction

Many complex problems can be modeled by using graphs, especially, when the underlying instances are net-like. The cells of a net usually correspond to nodes of a graph and the wires connecting cells to edges of this graph. If we aim to somehow cluster the cells under some conditions on wires connecting cells in different clusters then we obtain a graph partitioning problem. Despite of such an easy representation, problems of this kind are generally difficult to solve, i.e., even using state of the art computers an optimal solution for large instances cannot be achieved in a reasonable amount of time. Hence graph partitioning problems offer interesting and difficult mathematical questions.

The graph partitioning problem we consider in this thesis consists in clustering nodes of an edge- and node-weighted graph into subsets with restricted weight capacity so that the weighted sum of edges between nodes in different clusters is minimized. They arise e.g. in compiler design, parallel computing for scientific simulations and the placement problem in VLSI design.

The compiler design application is described e.g. in [53]. A compiler consists of several modules, where each module is a set of procedures or subroutines with a corresponding memory requirement. The modules are combined so that they form clusters whose size is restricted by storage capacity. Modules assigned to different clusters cause high communication costs, because the communication might require memory swapping. In the optimal compiler design the modules are to be assigned to clusters so that the storage requirements are met and the total communication costs between modules in different clusters are minimal. Representing the modules as nodes of a graph and the communications between modules as edges of this graph the problem transforms into a graph partitioning problem.

Another example: algorithms that find partitions of graphs are critical for the efficient execution of scientific simulations on high performance parallel computers, as reported for instance in [24, 80]. In these simulations computations are performed iteratively on each element of a mesh and then the information is exchanged between adjacent mesh elements. For example, computations are carried out on each triangle of the mesh shown in Figure 1.1. Then information is exchanged for every face between adjacent triangles.

The efficient execution of such simulations on parallel machines requires a mapping of the computational mesh onto processors such that each processor gets roughly an equal number of mesh elements and that the amount of inter-processor communication for the information exchange between adjacent elements is minimized. Such a mapping is commonly found by solving a graph partitioning problem.

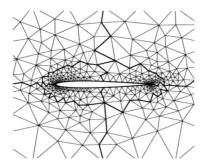

Figure 1.1: A triangular mesh surrounding an airfoil. The mesh is distributed among 8 processors, with divisions shown by bold lines.

The placement problem in VLSI design is considered for instance in [54, 64]. Given are components of a chip and a master divided into base cells, where the components are to be placed, as well as the list of nets, we want to find a placement of the components on the master without overlapping and yielding minimal costs in terms of minimum routing length, minimum number of vias, etc., see Figure 1.2. The problem can be tackled by a decomposition of the base cells into small number of clusters. It has to be decided to which cluster a given base cell has to be assigned to. The base cells correspond to nodes of a graph, their areas describe the weight of the nodes. The list of nets is translated to edges joining the nodes and their lengths are the costs. The more clusters, the higher the complexity of the problem due to the growing number of needed decision variables. Thus a common approach is to use bisection recursively: first the nodes are partitioned into two clusters and then again nodes in each of those clusters are partitioned in further two clusters etc. One set of data we use in the experimental part of this thesis originates in this application.

Graph partitioning problems belong to challenging topics in graph theory and combinatorial optimization. Due to their rich practical background many heuristic algorithms have been developed over the years to solve these problems. Such methods produce a solution quickly but do not guarantee optimality. Methods, which prove optimality of a solution or at least give good approximations of the optimal solution, are still time consuming. Nevertheless, recent developments allow to tackle instances of several thousands of nodes, a size intractable just a few years ago. These exact methods are rooted in combinatorial optimization. One mainstream approach uses integer programming methods,

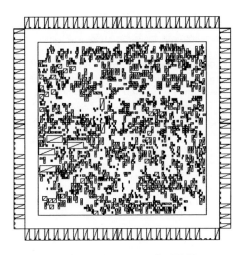

Figure 1.2: A cell placement of a VLSI circuit.

like branch-and-cut algorithm based on a linear relaxation. Another approach gaining on popularity is to apply semidefinite programming methods with spectral bundle algorithms.

The most successful method for solving integer linear programs follows the divide and conquer strategy called *branch-and-bound* introduced in 1960. These subproblems are created by restricting the range of integer variables. They are solved approximately using the *linear relaxation*, where integrality of the variables is neglected. The bounds delivered by the relaxation are compared and subproblems successively eliminated. If in the course of solving the linear relaxation additional, usually problem related, restrictions called *cutting planes* are added, the algorithm turns to a *branch-and-cut* method. The success of the cutting plane approach was released by Dantzig, Fulkerson, and Johnson in 1954 [23], who applied it to an instance with 42 cities (an impressive size at that time) of the *traveling salesman problem*, a classical combinatorial optimization problem. The recent best results in this area are achieved using branch-and-cut and tackle instances with 25 thousand locations and more. The success of the method for the traveling salesman problem inspired the researchers and practitioners to apply it to other combinatorial problems. Over the years it became the framework of the state of the art commercial and non-commercial solvers for integer programming problems.

In 1979 Lovász [67] introduced semidefinite programming to combinatorial optimization. Soon after several results supporting the strength of semidefinite relaxations followed. They led to a recent progress in approximation algorithms, initiated by the paper of Goemans and Williamson [38], confirming the importance of semidefinite programming for combinatorial optimization. With the development of efficient algorithms like interior point or spectral bundle methods semidefinite programming became an attractive alternative to linear programming for approximations of combinatorial problems like graph partitioning problems. The next step in utilization of the semidefinite methods was the integration of cutting plane algorithms developed by Helmberg [44]. Soon after the approach to combine a semidefinite solver with branch-and-cut followed. In [86, 77] Rendl, Rinaldi, and Wiegele apply the interior point method to solve the semidefinite relaxation. The novelty in our approach is to use the spectral bundle method. The first results can be found in the PhD thesis of Armbruster [4], our project partner. Further investigations are published within this work.

The *minimum graph bisection problem*, the special graph partitioning problem being the subject of our study, concerns a partition of the node set of a graph with weighted nodes and edges into two subsets with bounded total node weight so that the weighted sum of edges joining nodes in different clusters is minimal. The problem can be modeled by means of integer linear as well as integer quadratic programming. Both formulations naturally lead to a linear and a semidefinite relaxation, respectively. We investigate the quality of these relaxations in a branch-and-cut algorithm. To accelerate the solution process the relaxations are improved by studying the *bisection cut polytope* associated with the underlying problem.

We give now a detailed summary on the content of the thesis. In Chapter 2 the minimum graph bisection problem is introduced together with integer linear and quadratic programming formulations and corresponding relaxations. Furthermore, we give a survey on the literature concerning polyhedral studies associated with graph partitioning problems, and we put together known results about the bisection cut polytope.

In the next three chapters we continue studying the bisection cut polytope. We apply and extend known results for related graph partitioning problems and associated polytopes to our graph bisection case. In Chapter 3 we consider *tree and cycle inequalities* which we upgrade to *complementarity tree* and *cycles with trees inequalities* by taking advantage of the fact that the nodes of the graph are divided into exactly two sets.

In 1996 Ferreira et al. introduced the so-called *knapsack tree inequalities* which involve the knapsack conditions on the nodes and turned out to be computationally very efficient. In Chapter 4 we give sufficient and necessary conditions for the knapsack tree inequality to be facet-defining. This result substantiates the computational success of these inequalities.

In Chapter 5 we generalize the knapsack tree inequalities to *bisection knapsack walk inequalities* by exploiting again the fact that the nodes of the graph are partitioned into exactly two subsets. The class of inequalities called *capacity reduced bisection knapsack walk inequalities* extends both classes of inequalities to non-connected substructures. The idea is to exploit the weights of nodes that are not included in the subgraph supporting the corresponding knapsack inequality to reduce the right-hand sides. These stronger conditions result in nonlinear right-hand sides. We consider the convex envelope of this nonlinear function and show that the supporting hyperplanes are in one-to-one correspondence to the faces of a certain polytope, called *cluster weight polytope*. For a special case of the underling graph a complete description of the cluster weight polytope is established. This yields the tightest strengthening possible for the capacity reduced bisection knapsack walk inequalities. The main contribution to this topic is due to Armbruster and can be found in his PhD thesis [4]. We present an alternative proof for the complete description of the cluster weight polytope. Our proof bases on the observation that the coefficients of the facets of the cluster weight polytope can be established by solving a continuous linear knapsack problem (see Section A.5).

The efficiency of a branch-and-cut algorithm strongly depends on the quality of the bounds delivered by the dual and primal methods yielding in our case lower and upper bounds, respectively. Good bounds contribute to the reduction of the size of the branch-and-bound tree. The polyhedral investigations are responsible for tightening the lower bound. In Chapter 6 we present heuristic methods which exploit solutions of the relaxations to construct upper bounds. Here we pursuit randomized methods like genetic algorithms or greedy randomized adaptive search procedures.

Finally, we devote Chapter 7 to empirical investigations. The theoretical results from previous chapters are integrated into a branch-and-cut framework, equipped with solvers for the linear and semidefinite relaxation, in form of separation algorithms and primal heuristics. We show the impact of the polyhedral studies on both relaxations using instances coming from VLSI design and scientific computations as well as random graphs. Generally only the linear relaxation benefits from the new inequalities. Although they also improve the bounds delivered by semidefinite relaxation, the separation time significantly slows down the solution process. The semidefinite relaxation seems to outperform the linear one due to a weakness of the primal methods applied to the linear relaxation. Nevertheless, very dense instances (randomly generated) are still solved more efficiently using the linear relaxation in comparison to the semidefinite one.

The mathematical models and methods presented in this thesis have their roots in graph theory and polyhedral theory as wells as linear, integer and semidefinite programming. In the appendix we introduce selected notions from these topics which should facilitate the reader to follow the content of this thesis in detail. We also supply the reader with references, where the outlined subjects are treated more profoundly. On page 206 we list frequently used notations.

Chapter 2

The Minimum Graph Bisection Problem

The minimum graph bisection problem once defined, we give a survey on contributions in the literature related to graph partitioning problems which set up our research. Next, we present several equivalent integer programming formulations for the minimum graph bisection problem and the corresponding linear and semidefinite relaxations. The feasible sets of these programs define the bisection cut polytope which plays a significant role throughout this work.

2.1 Problem Definition

For convenience we begin with a few definitions and notations. For full details we refer the reader to Chapter A.

We consider a simple, undirected and weighted graph $G = (V, E)$, where V is the node set and E is the edge set of G. To each node $v \in V$ the weight $f_i \in \mathbb{Z}_+$ and to each edge $e \in E$ the cost $w_e \in \mathbb{R}_+$ are assigned.

For a, possibly empty, set $S \subseteq V$ the **cut** $\Delta(S)$ induced by S is the set of edges having one end-node in S and one end-node in $V \setminus S$,

$$\Delta(S) = \{uv \in E : u \in S, v \in V \setminus S\}.$$

A **partition** $(V_1, \ldots, V_k)$ of the node set V is a collection of mutually disjoint sets $V_i \subseteq V$, $i = 1, \ldots, k$, such that $V = V_1 \cup \ldots \cup V_k$. If $V_i \neq \emptyset$ for each $i = 1, \ldots k$ then the sets $V_1, \ldots, V_k$ are called **clusters** and the partition $(V_1, \ldots, V_k)$ is called a **cluster partition**. The edge set

$$\Delta(V_1, \ldots, V_k) = \{uv \in E : u \in V_i, v \in V_j, i \neq j\}$$

7

is called **multicut** or **k-cut**. For $k = 2$ the partition (V_1, V_2) is called **bipartition** and $\Delta(V_1, V_2) = \Delta(V_1) = \Delta(V_2)$.

Let A and B be two sets such that $A \subseteq B$. A vector $\chi^A \in \{0, 1\}^{|B|}$ such that

$$\chi_i^A = \begin{cases} 1, & i \in A, \\ 0, & i \in B \setminus A, \end{cases}$$

is called the **incidence vector of** A. For a vector $v \in \mathbb{R}^n$ and a subset of indices I we abbreviate

$$v(I) := \sum_{i \in I} v_i.$$

For a given real number $\tau \in [0, 1]$ we define the **upper bisection bound**

$$u_\tau = \left\lceil \frac{1 + \tau}{2} f(V) \right\rceil.$$

The number τ is referred to as **bisection ratio**. A bipartition $(S, V \setminus S)$ such that the total node weight of each cluster is less or equal u_τ, i.e.,

$$f(S) \le u_\tau \quad \text{and} \quad f(V \setminus S) \le u_\tau \tag{2.1}$$

holds, is called a **bisection**. A cut $\Delta(S)$ corresponding to a bisection $(S, V \setminus S)$ is called **bisection cut**. The **minimum graph bisection problem** (MGBP) is to find a bisection $(S, V \setminus S)$ with the minimum cost bisection cut,

(MGBP) $\min \{ w(\Delta(S)) : S \subseteq V, f(S) \le u_\tau \text{ and } f(V \setminus S) \le u_\tau \}.$

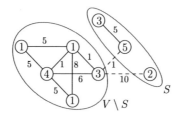

Figure 2.1: An example of a graph bisection $(S, V \setminus S)$ with $\tau = 0.1$, $l_\tau = 9$ and $u_\tau = 11$. The dashed edges belong to the bisection cut $\Delta(S)$. The numbers within the nodes and the numbers at the edges correspond to node and edge weights respectively.

The MGBP is known to be NP-hard [36][1] (problem reference number ND14) as the Graph Partitioning Problem, see Problem A.17.

---

[1]This reference contains explanation to all terms we use concerning the complexity theory.

The convex hull of all incidence vectors of bisection cuts with respect to the set E,

$$P_{\mathcal{B}} := \text{conv}\{y \in \{0,1\}^{|E|} : y = \chi^{\Delta(S)}, S \subseteq V, (S, V \setminus S) \text{ is a bisection in } G\},$$

is called **bisection cut polytope**.

For convenience we define the **lower bisection bound**

$$l_\tau := f(V) - u_\tau.$$

Note that (2.1) is then equivalent to

$$f(S) \geq l_\tau \quad \text{and} \quad f(V \setminus S) \geq l_\tau.$$

A set of nodes $S \subseteq V$ with $f(S) > u_\tau$ is called **cover**.

2.2 Polyhedral Studies - Survey on Literature

Despite the wide range of heuristic methods developed for the graph partitioning problems, see Section 6.1, exact methods are in demand. The usual approach here is to formulate the problem as a binary or integer program and investigate the associated polytope defined by the convex hull of the feasible solutions in order to obtain its facial description. Next, we present results on polyhedral studies concerning problems strongly related to the minimum graph bisection problem. All mentioned inequality classes are defined in Section 2.5 below.

If $\tau = 1$ the bisection cut polytope $P_{\mathcal{B}}$ coincides with the so called **cut polytope** $P_{\mathcal{C}}$. A profound study of this polytope, composed by Deza and Laurent in [27], arose in connection with the **max-cut problem**

$$\max\{w(\Delta(S)) : S \subseteq V, \Delta(S) \subseteq E \text{ is a cut in } G = (V, E)\}.$$

Since $P_{\mathcal{C}}$ is the convex hull of incidence vectors of all possible cuts in G, it is the fundamental polytope for most partitioning problems. Barahona and Mahjoub [11] introduce a class of valid inequalities for $P_{\mathcal{C}}$ called odd cycle inequalities together with a polynomial time algorithm for their separation which we describe in Section 3.4. They show the strength of these inequalities and describe cases, when these inequalities fully describe $P_{\mathcal{C}}$. Alevras [3] gives a full facial description of cut polyhedra for complete graphs up to seven nodes. Among the identified facet-defining inequalities are tree, cycle, and cycle with tails inequalities. De Simone and Rinaldi [82] solve the max-cut problem on complete graphs by applying a cutting plane algorithm for the hypermetric inequalities which generalize the triangle inequalities for complete subgraphs with number of nodes greater than 3.

A popular partitioning problem called **equipartition, equicut** or **2-partition** problem concerns dividing the node set of a graph into two subsets with possibly equal number of nodes. It can be obtained from the MGBP by taking unique node weights and setting $\tau = 0$. The polytope associated with the equicut problem,

$$P_{\mathcal{E}} := \mathrm{conv}\{ y \in \{0,1\}^{|E|} : y = \chi^{\Delta(S)},\ S \subseteq V,\ |S| = |V \setminus S|\ \text{or}\ |S| = |V \setminus S| \pm 1 \},$$

is well-investigated by Conforti, Rao, and Sassano in [19, 20]. The authors give results concerning the dimension as well as the facial structure of the polytope including a wide range of valid inequalities for $P_{\mathcal{E}}$. Among them are star, tree, and cycle inequalities. In [24, 25] de Souza and Laurent further investigate the equicut polytope and give generalizations of tree and cycle inequalities.

In [17] Chopra and Rao focus on a polyhedral approach to solve **k-partition** problems which bears on partitioning the nodes into at most, at least or exactly k clusters by minimizing the total weight of edges with end-nodes in different subsets. Among the defined inequalities are specializations of clique, tree, and cycle inequalities.

The **node capacitated graph partitioning problem** NCGPP, which arises in the VLSI design application described in Chapter 1, is investigated by Ferreira, Martin, de Souza, Weismantel, and Wolsey in [31]. This problem concerns the partition of a weighted node and edge graph into $K (\geq 2)$ clusters, so that the total node weight of each cluster does not exceed a given bound F and the total weight of edges joining nodes in different subsets is minimized. The NCGP turns to the MGBP if we set $K = 2$ and $F = u_\tau$. Hence the minimum graph bisection problem can be seen as a special case of NCGPP. Due to this fact almost all results derived for the polytope $P_{\mathcal{F}}$,

$$P_{\mathcal{F}} := \mathrm{conv}\{ y \in \{0,1\}^{|E|} : y = \chi^{\Delta(V_1,\dots,V_K)},\ f(V_i) \leq F,\ i = 1,\dots,K \},$$

associated with NCGPP are valid for $P_{\mathcal{B}}$. In [31] known valid inequalities for the cut and equicut polytope, mentioned above, are adjusted for $P_{\mathcal{F}}$ and new classes of inequalities like knapsack tree or generalizations of cycle inequalities are introduced.

In [48] Holm and Sørsen model NCGPP by maximizing the weighted sum of edges between nodes in the same subsets. They also consider inequalities based on the knapsack condition.

We discuss the computational success of the polyhedral approach for solving the listed graph partitioning problems in Section 7.1.

Known Valid Inequalities

The large variety of valid inequalities for the polytopes $P_{\mathcal{C}}$, $P_{\mathcal{E}}$, and $P_{\mathcal{F}}$, defined above, can be applied or adapted for the bisection cut polytope $P_{\mathcal{B}}$ due to its inclusion in or intersection with these polytopes. We present below original definitions of those classes of inequalities which we will work on in the course of this thesis.

Theorem 2.1 (odd cycle inequality, Barahona, Mahjoub, 1986) *Let the subgraph* $C = (V_C, E_C)$ *be a cycle in* G. *Let* D *be a subset of* E_C *such that* $|D|$ *is odd. Then*

$$\sum_{e \in D} y_e - \sum_{e \in E_C \setminus D} y_e \leq |D| - 1$$

is a valid inequality for the cut polytope P_C.

If the cycle in the above definition reduces to a triangle ($V_C = \{i, j, k\}$) then we obtain the well-known **triangle inequalities**,

$$y_{ij} + y_{jk} - y_{ik} \geq 0,$$
$$y_{ik} + y_{kj} - y_{ij} \geq 0,$$
$$y_{ij} + y_{ik} - y_{jk} \geq 0,$$
$$y_{ij} + y_{ik} + y_{jk} \leq 2.$$

As we already mentioned, in [19, 20] Conforti et al. introduce valid inequalities for the polytope $P_{\mathcal{E}}$ associated with the equipartition problem. Below we cite inequalities which will be adapted for the bisection case in Chapter 3.

Theorem 2.2 (tree inequality, Conforti et al., 1990) *Let* $T = (V_T, E_T)$ *be a subtree of the graph* $G = (V, E)$ *with* $|V| = 2p + 1$ *or* $|V| = 2p$. *If* $|V_T| > p + 1$ *then*

$$\sum_{e \in E_T} y_e \geq 1$$

is valid for $P_{\mathcal{E}}$.

Theorem 2.3 (star inequalities, Conforti et al., 1990) *Given a graph* $G = (V, E)$ *and a node* $r \in V$ *adjacent to all other nodes in* G, *the inequalities*

$$\sum_{e \in \Delta(\{r\})} y_e \geq \left\lfloor \frac{|V|}{2} \right\rfloor,$$

$$\sum_{e \in \Delta(\{r\})} y_e \leq \left\lfloor \frac{|V|}{2} \right\rfloor + 1$$

are valid for $P_{\mathcal{E}}$.

Theorem 2.4 (cycle inequality, Conforti et al., 1990) *Let* $C = (V_C, E_C)$ *be a subcycle of* $G = (V, E)$ *with* $|V| = 2p + 1$ *or* $|V| = 2p$. *If* $|V_C| > p + 1$ *then*

$$\sum_{e \in E_C} y_e \geq 2$$

is valid for $P_{\mathcal{E}}$.

Adequate inequalities for the polytope $P_{\mathcal{F}}$ are presented in [31] by Ferreira et al.

Theorem 2.5 (tree inequality, Ferreira et al., 1996) *Let* $T = (V_T, E_T)$ *be a subtree of* $G = (V, E)$. *If* $f(V_T) > F$ *then*

$$\sum_{e \in E_T} y_e \geq 1$$

is valid for $P_{\mathcal{F}}$.

Theorem 2.6 (star inequality, Ferreira et al., 1996) *Let* (V_S, E_S) *be a substar of* $G = (V, E)$ *centered at a node* $r \in V_S$. *If* $|\Delta(\{r\})| \geq F$, *then the inequality*

$$\sum_{e \in \Delta(\{r\})} y_e \geq |\Delta(\{r\})| - F + 1 \tag{2.2}$$

is valid for $P_{\mathcal{F}}$.

Theorem 2.7 (cycle inequality, Ferreira et al., 1996) *Let* $C = (V_C, E_C)$ *be a subcycle of* $G = (V, E)$. *If* $f(V_C) > F$ *then*

$$\sum_{e \in E_C} y_e \geq 2 \tag{2.3}$$

is valid for $P_{\mathcal{F}}$.

Theorem 2.8 (cycle with tails inequality, Ferreira et al., 1996) *Let the subgraph* $G' = (V', E')$ *of* G *be a cycle with tails, i.e., there exists a cycle* $C = (V_C, E_C)$, $V_C \subseteq V'$, *and paths* $(V_1, E_1), \dots, (V_p, E_p)$ *such that* $|V_i \cap V_C| = 1$, $V_i \cap V_j \cap (V' \setminus V_C) = \emptyset$, $i, j \in \{1, \dots, p\}$, *and* $V' = V_C \cup V_1 \cup \dots V_p$, $E' = E_C \cup E_1 \cup \dots E_p$. *If* $f(V') > F$ *then*

$$\sum_{e \in E_C} y_e + 2 \sum_{i=1}^{p} \sum_{e \in E_i} y_e \geq 2$$

is valid for $P_{\mathcal{F}}$.

Another type of inequalities introduced in [31] makes more specific use of the node weights f_i, $i \in V$, and cluster capacity F. In fact, valid inequalities for the knapsack polytope,

$$\text{conv}\{x \in \{0, 1\}^{|V|} : \sum_{v \in V} f_v x_v \leq F\}, \tag{2.4}$$

related to the constraint restricting the cluster capacities are applied.

Theorem 2.9 (knapsack tree inequality, Ferreira et al., 1996) *Let*

$$\sum_{v \in V} a_v x_v \leq a_0,$$

with $a_v \geq 0$ for all $v \in V$, be a valid inequality for the knapsack polytope (2.4). Select a node $r \in V$ and a subtree $T = (V_T, E_T)$ of G rooted at r. For each $v \in V_T \setminus \{r\}$ let (V_{rv}, E_{rv}) be the path in T joining v with the root r. Then the inequality

$$\sum_{v \in V_T} a_v \left(1 - \sum_{e \in E_{rv}} y_e \right) \leq a_0$$

is valid for $P_{\mathcal{F}}$.

This inequality adapted for the bisection case is treated in details in Chapters 4 and 5.

For convenience we introduce now the following definitions. The polytope associated with the constraint on cluster capacity limits (2.1),

$$P_{\mathcal{K}} = \mathrm{conv}\{x \in \{0,1\}^{|V|} : \sum_{v \in V} f_v x_v \leq u_\tau\},$$

will be called the **knapsack polytope**. We refer also to Section A.5 for further details on the associated knapsack problem. Given a valid inequality

$$\sum_{e \in E} \alpha_e y_e \leq \alpha_0$$

for $P_{\mathcal{B}}$. The set of edges $\overline{E} := \{e \in E : \alpha_e \neq 0\}$ is called the **support** of $\alpha y \leq \alpha_0$ and the graph $(V(\overline{E}), \overline{E})$ induced by $\overline{E}$ is the **subgraph** of G **supporting** this inequality.

2.3 Linear Integer Programming Models and Linear Relaxations

For an integer programming formulation of the minimum graph bisection problem we introduce binary variables y_{ij} for all $ij \in E$. Each $y \in \{0,1\}^{|E|}$ satisfying $y_{ij} = 1$ if nodes i and j are in different clusters, and $y_{ij} = 0$ otherwise, corresponds to an incidence vector of a cut in G. Let

$$Y := \{y \in \{0,1\}^{|E|} : y = \chi^{\Delta(S)}, \, S \subseteq V, \, (S, V \setminus S) \text{ is a bisection in } G \}.$$

We seek for such an element of set Y that minimizes $\sum_{e \in E} w_e y_e$. It remains to describe Y by linear constraints. One formulation can be derived directly from the model for NCGPP, presented in [31], by setting the number of clusters to 2. Introducing variables z_i^k for each node $i \in V$ and each cluster $k = 1, 2$ we obtain

$$\min \sum_{e \in E} w_e y_e,$$

$$\begin{aligned}
\text{s.t.} \quad z_i^1 + z_i^2 &= 1, & \forall i \in V, \\
z_i^1 - z_j^1 &\leq y_{ij}, & \forall ij \in E, \\
z_j^1 - z_i^1 &\leq y_{ij}, & \forall ij \in E, \\
2 - z_i^1 - z_j^1 &\geq y_{ij}, & \forall ij \in E, \\
2 - z_j^2 - z_i^2 &\geq y_{ij}, & \forall ij \in E, \\
\sum_{i \in V} f_i z_i^k &\leq u_\tau, & k = 1, 2 \\
y_{ij} &\in \{0, 1\}, & \forall ij \in E, \\
z_i^1, z_i^2 &\in \{0, 1\}, & \forall i \in V.
\end{aligned}$$

(IP1)

The first constraint ensures that each node belongs to exactly one cluster. The next four inequalities guarantee that $y_{ij} = 1$ if and only if $z_i^1 \neq z_j^1$, i.e., nodes i and j are in different clusters, and $y_{ij} = 0$ otherwise. In particular, the first pair of these constraints forces y_{ij} to take the value 1 if $z_i^1 \neq z_j^1$. The second pair sets y_{ij} to 0 if both nodes i and j are in one of the two clusters. The last constraint ensures that the total weight of nodes in each cluster stays within the lower and the upper bound[2]. Due to our formulation each solution $y \in \{0, 1\}^{|E|}$ of the above problem corresponds to an incidence vector of a bisection cut and vice versa, for all bisection cuts their incidence vectors are solutions of (IP1). Hence the projection of the feasible set of (IP1) onto y-space equals set Y.

Due to the fact that in our partitioning problem we consider only two clusters we may simplify the above model by reducing the number of variables. In fact we need only one binary variable z_i for each node $i \in V$ and require that all z-variables corresponding to nodes assigned to one cluster have the same value, see Figure 2.2.

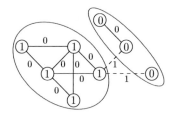

Figure 2.2: Values of variables y_{ij}, $ij \in E$, and z_i, $i \in V$, corresponding to the bisection presented in Figure 2.1.

Then the following constraints guarantee that $y_{ij} = 1$ if and only if $z_i \neq z_j$, i.e., nodes i

---

[2]Substituting $z_i^1 = 1 - z_i^2$ for $k = 1$ and $z_i^2 = 1 - z_i^1$ for $k = 2$, $i \in V$, and using the relation $l_\tau = f(V) - u_\tau$ one obtains $\sum_{i \in V} f_i z_i^k \geq l_\tau$, $k = 1, 2$.

and j are in different clusters, and $y_{ij} = 0$ otherwise:

$$z_i - z_j \leq y_{ij},$$

$$z_j - z_i \leq y_{ij},$$

$$z_j + z_i \geq y_{ij},$$

$$z_i + z_j \leq 2 - y_{ij}.$$

The constraints restricting the capacities of node clusters take the form

$$\sum_{i \in V} f_i z_i \leq u_\tau,$$

$$\sum_{i \in V} f_i (1 - z_i) \leq u_\tau.$$

Simplified by the relation $l_\tau = \sum_{i \in V} f_i - u_\tau$ they both reduce to

$$l_\tau \leq \sum_{i \in V} f_i z_i \leq u_\tau. \tag{2.5}$$

Summing up we obtain the second integer programming formulation for the MGBP:

$$\min \sum_{e \in E} w_e y_e,$$

$$
\begin{array}{llll}
\text{s.t.} & z_i - z_j & \leq & y_{ij}, & \forall\, ij \in E, \\
& z_j - z_i & \leq & y_{ij}, & \forall\, ij \in E, \\
& z_j + z_i & \geq & y_{ij}, & \forall\, ij \in E, \\
(\text{IP2}) & z_i + z_j & \leq & 2 - y_{ij}, & \forall\, ij \in E, \\
& l_\tau & \leq & \sum_{i \in V} f_i z_i \; \leq \; u_\tau, & \\
& y_{ij} & \in & \{0, 1\}, & \forall\, ij \in E, \\
& z_i & \in & \{0, 1\}, & \forall\, i \in V.
\end{array}
$$

Again the role of the z-variables is to force the values of y-variables so that the resulting vector corresponds to an incidence vector of some bisection cut. Hence the projection of the feasible set defined in (IP2) onto the y-space equals set Y. Note that, since our objective is to minimize a positive weighted sum of y-components, the constraint

$$z_i + z_j \leq 2 - y_{ij}, \qquad \forall\, ij \in E,$$

is redundant in the problem formulation.

Since the node variables z_i, $i \in V$, do not appear in the objective function, one can be tempted to get rid of them. We do it in the following way. We select a node $s \in V$ and extend the set of edges E so that s is adjacent to all other nodes in V. The weights w_{is} of new edges are set to zero, see Figure 2.3.

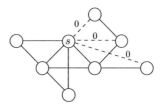

Figure 2.3: New edges (dashed) incident to node s and their costs.

W.l.o.g. we assume that $z_s = 1$. Then for each node i in V we have $z_i = 1 - y_{is}$. Using this transformation we reformulate next the constraint in (IP2) restricting the total node weight in the clusters. For all nodes $i \in V$ which are in the same cluster as s there holds

$$l_\tau \leq f_s + \sum_{i \in V \setminus \{s\}} f_i(1 - y_{is}) \leq u_\tau. \tag{2.6}$$

For the other cluster we have

$$l_\tau \leq \sum_{i \in V \setminus \{s\}} f_i y_{is} \leq u_\tau. \tag{2.7}$$

Because of $l_\tau + u_\tau = \sum_{i \in V} f_i$ (2.6) is equivalent to (2.7). As the example in Figure 2.4 shows, we need now to ensure that each cycle in G contains an even number of edges from any cut.

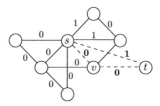

Figure 2.4: An optimal solution to the problem $\min\{ w^T y \: : \: y \in \{0, 1\}^{|E|}$ and (2.7) $\}$ for the graph presented in Figure 2.1. Due to $y_{st} = 1$ nodes s and t should be in different clusters, because of $y_{vt} = y_{st} = 0$ nodes s and t should be in the same cluster.

For that we apply the **odd cycle inequalities**[3] of Barahona and Mahjoub [11],

$$\sum_{e \in D} y_e - \sum_{e \in C \setminus D} y_e \leq |D| - 1, \quad \forall C \subseteq E, \: D \subseteq C, \: |D| \text{ is odd}, \tag{2.8}$$

---

[3]See also Theorem 2.1 in Section 2.5 as well as Section 3.4.

where $(V(C), C)$ is a cycle in G. Thus we obtain the third integer programming formulation for the MGBP:

$$\min \sum_{e \in E} w_e y_e,$$

(IP3)

$$\text{s.t.} \quad (2.7), (2.8),$$

$$y_e \in \{0, 1\}, \, \forall e \in E.$$

In this formulation constraint (2.8) guarantees that each solution corresponds to an incidence vector of some cut in G and (2.7) ensures that this cut is a bisection cut. Hence the feasible solutions to (IP3) are in one-to-one correspondence to the elements of Y.

Linear relaxations of all integer programs presented above arise naturally by extending the domain of y- and z-variables from the discrete set $\{0, 1\}$ to the continuous interval $[0, 1]$.

2.4 Quadratic Integer Programming Model and Semidefinite Relaxation

The MGBP can be also modeled as an integer quadratic program. Such a formulation leads to a semidefite relaxation of the MGBP. We follow the ansatz of Helmberg given in [42] when deriving the semidefinite program. Given a bisection $(S, V \setminus S)$ we introduce variables $x_i \in \{-1, 1\}$ for each $i \in V$ and demand that

$$x_i = \left\{ \begin{array}{ll} 1, & i \in S, \\ -1, & i \in V \setminus S. \end{array} \right.$$

Using the node-variables transformation

$$z_i = \frac{x_i + 1}{2}, \quad i \in V,$$

as well as the relation $l_\tau = \sum_{i \in V} f_i - u_\tau$ we apply the cluster capacity restriction (2.5) to x-variables and obtain

$$l_\tau - u_\tau \leq \sum_{i \in V} f_i x_i \leq u_\tau - l_\tau,$$

i.e.,

$$|\sum_{i \in V} f_i x_i| \leq u_\tau - l_\tau. \tag{2.9}$$

Now, using the transformation

$$y_{ij} = \frac{1 - x_i x_j}{2}, \quad ij \in E, \tag{2.10}$$

we obtain $y_{ij} = 0$ if nodes i and j are in one cluster, and 1 otherwise, see Figure 2.5.

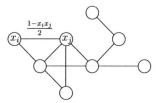

Figure 2.5: x-variables for the quadratic programming formulation (QP).

Thus the quadratic formulation for the MGBP reads:

$$\min \sum_{ij \in E} w_{ij} \frac{1 - x_i x_j}{2},$$

(QP)

$$\text{s.t.} \quad \left| \sum_{i \in V} f_i x_i \right| \le u_\tau - l_\tau,$$

$$x_i \in \{-1, 1\}, \, \forall \, i \in V.$$

Exploiting the symmetry of the weighted adjacency matrix W of the graph G and the fact that $x_i x_i = 1$, $i \in V$, we modify the objective function in (QP) as follows:

$$\sum_{ij \in E} w_{ij} \frac{1 - x_i x_j}{2} = \sum_{i<j} w_{ij} \frac{1 - x_i x_j}{2}$$

$$= \frac{1}{4} \sum_{1 \le i,j \le |V|} w_{ij}(1 - x_i x_j) \tag{2.11}$$

$$= \frac{1}{4} x^T (\text{Diag}(We) - W)x,$$

where $e = (1, \ldots, 1)^T$. $\text{Diag}(We) - W$ is the Laplace matrix of the graph G. We consider next the matrix $X := xx^T$. It is a positive semidefinite matrix with rank(X)=1, due to its definition, and with diagonal entries equal to 1, because of our assumptions on vector x. Setting

$$C := \tfrac{1}{4}(\text{Diag}(We) - W)$$

and applying the properties of the matrix's scalar product we transform (2.11) further to

$$\sum_{ij \in E} w_{ij} \frac{1 - x_i x_j}{2} = x^T C x = \langle Cx, x \rangle = \langle C, xx^T \rangle = \langle C, X \rangle$$

as well as (2.9) to

$$(u_\tau - l_\tau)^2 \geq |f^T x|^2 = \langle f f^T, X \rangle. \tag{2.12}$$

Then the following formulation of the MGBP is equivalent to (QP):

$$\min \langle C, X \rangle,$$

(QP2)
$$\text{s.t.} \quad \langle f f^T, X \rangle \leq (u_\tau - l_\tau)^2,$$

$$\text{diag}(X) = e,$$

$$\text{rank}(X) = 1,$$

$$X \succeq 0,$$

see also [63]. Dropping the "rank one constraint", which disrupts the convexity of the feasible set of (QP2), yields a semidefinite programming relaxation of (QP2) and hence of (QP),

$$\min \langle C, X \rangle,$$

(SP)
$$\text{s.t.} \quad \langle f f^T, X \rangle \leq (u_\tau - l_\tau)^2,$$

$$\text{diag}(X) = e,$$

$$X \succeq 0.$$

2.5 Bisection Cut Polytope

In [81] De Simone shows the equivalence between the cut polytope and the boolean quadratic polytope, associated with the quadratic formulation of the max-cut problem, under some affine transformation. We derive now a similar equivalence between the bisection cut polytopes associated with the integer quadratic and the integer linear programming formulations for the minimum graph bisection problem.

Consider the feasible set of (QP) and (QP2),

$$\mathcal{X} = \{ xx^T : x \in \{-1, 1\}^{|V|}, |f^T x| \leq u_\tau - l_\tau \}.$$

For any $X \in \mathcal{X}$ the entries X_{ij}, $ij \in E$, deliver the components of the incidence vector of a bisection cut y by means of the affine transformation (2.10),

$$y_{ij} = \frac{1 - X_{ij}}{2}, \quad ij \in E. \tag{2.13}$$

Hence

$$\text{conv}\{[\tfrac{1-X_{ij}}{2}]_{ij \in E} : X \in \mathcal{X}\} = \{[\tfrac{1-X_{ij}}{2}]_{ij \in E} : X \in \text{conv}(\mathcal{X})\}$$

defines the bisection cut polytope $P_{\mathcal{B}}$. On the other hand, following the definition of set Y, considered in Section 2.3, we have

$$P_{\mathcal{B}} = \text{conv}(Y),$$

and hence $P_{\mathcal{B}}$ is naturally included in the feasible sets of linear relaxations of all integer programs presented in Section 2.3. Thus both linear and semidefinite relaxations can be strengthened by valid inequalities for $P_{\mathcal{B}}$.

Dimension of the Bisection Cut Polytope

In this section we generally assume that $f_i = 1$ and we give some remarks on the dimension of the polytope $P_{\mathcal{B}}$ in this special case.

Proposition 2.10 (Ferreira et al. 1996) *For a graph $G = (V, E)$ with $u_\tau \geq 2$ and $|V| \leq 2u_\tau - 2$ the bisection cut polytope $P_{\mathcal{B}}$ is full-dimensional.*

(See Corollary 4.2 in [31].)

Due to the fact that

$$u_\tau = \frac{1 + \tau}{2} \sum_{i \in V} f_i = \frac{1 + \tau}{2} |V|$$

we have

$$|V| \leq 2u_\tau - 2 = (1 + \tau)|V| - 2.$$

Hence due to Proposition 2.10 $P_{\mathcal{B}}$ is full-dimensional if $\tau|V| \geq 2$. To discuss the remaining case, i.e., $0 \leq \tau|V| < 2$, we use the relation

$$u_\tau - l_\tau = \frac{1 + \tau}{2}|V| - \frac{1 - \tau}{2}|V| = \tau|V|.$$

The value $\tau|V|$ gives the difference between the total node weights of the two clusters. Since all nodes have unique (integer) weights, we consider only cases when $\tau|V|$ is an integer less than 2. Thus we obtain that either $\tau = 0$ or $\tau = \frac{1}{|V|}$. These cases specialize the equipartition problem and the corresponding equicut polytope extensively studied in [19].

Lemma 2.11 (Conforti et al. 1990) *Let $G = (V, E)$ be a graph with an odd number of nodes and let $\tau = \frac{1}{|V|}$. The polytope $P_\mathcal{B}$ is full-dimensional if and only if G is not a complete graph.*

(See Lemma 3.6 in [19].)

Lemma 2.12 (Conforti et al. 1990) *Let $G = (V, E)$ be a graph with an even number of nodes, i.e., $|V| = 2p$ for some $p \in \mathbb{N}$, and let $\tau = 0$. Consider a graph $\overline{G}$ which is a partial graph of K_{2p} induced by $\overline{E} = E(K_{2p}) \setminus E(G)$. The polytope $P_\mathcal{B}$ is full-dimensional if and only if $\overline{G}$ does not contain a bipartite connected component.*

(See Lemma 3.7 in [19].)

Summing up the above considerations we obtain the following assertions about the dimension of $P_\mathcal{B}$ if the graph G is a tree. We will need this result in Chapter 4.

Corollary 2.13 *Let the graph G be a tree and $|V| \geq 3$.*

(1) $P_\mathcal{B}$ *is full-dimensional for* $\tau \in (0, 1]$.

(2) *If* $\tau = 0$ *then* $P_\mathcal{B}$ *is full-dimensional if and only if G is not a star G and contains at least three mutually non-adjacent nodes.*

Proof. By Proposition 2.10 $P_\mathcal{B}$ is full-dimensional for $\tau \geq \frac{2}{|V|}$ and by Lemma 2.11 for $\tau = \frac{1}{|V|}$, since G as a tree is not a complete graph. Now, consider the case $\tau = 0$. If G is a star, then the complement graph $\overline{G}$ defined in Lemma 2.12 contains an isolated node which is a trivial bipartite component of $\overline{G}$. If G is not a star then $\overline{G}$ is a connected graph. If G contains at least three mutually non-adjacent nodes then $\overline{G}$ contains a triangle and thus it cannot be bipartite connected. $\qquad\square$

Chapter 3

Inequalities Depending on the Graph Structure

In this chapter we present several classes of valid inequalities for P_B, whose supports contain basic connected subgraphs of the given graph G like trees, cycles or their combinations, e.g. cycle with tails. The main idea is to find a connected subgraph, whose total node weight exceeds the limit u_τ, and in this way to force some edges of the subgraph to be cut by an arbitrary bisection in G. Depending on the structure of the subgraph one can estimate the minimum number of edges which must be in the cut. Some of such structure dependent inequalities like tree, cycle, and cycle with tails inequalities were already introduced for the polytopes $P_\mathcal{E}$ and $P_\mathcal{F}$. We extend these classes to **cycles with trees** inequalities as well as specify cases, when the right-hand side of the inequality can be increased.

The underlying bisection case gives us a comfortable situation in the sense that we deal only with two clusters of nodes. This simplifies dealing with connected subgraphs containing edges which are supposed to be in the cut. Moreover, we can specify new cases that lead to the overweight of one of the cluster. In this way we obtain **complementarity tree** and **complementarity cycles with trees** inequalities.

The odd cycle inequalities, which are part of the integer programming formulation (IP3), belong also to the class of structure dependent inequalities. For the sake of completeness we finish this chapter with the validity proof of this inequality class for the bisection cut polytope.

For all mentioned classes of inequalities we give separation algorithms which we incorporate in our branch-and-cut framework.

Ferreira et al. presented in [31] several classes of valid inequalities for $P_\mathcal{F}$ depending on the graph structure like star, tree, cycle, or cycle with tails inequalities, see Section 2.2. They were able to show that these inequalities define facets of $P_\mathcal{F}$ in very special cases.

For instance, the star inequality (2.6) and the cycle inequality (2.7) are facet defining for $P_{\mathcal{F}}$ if the node set is partitioned in as many clusters as nodes in the graph.

Studying full descriptions of small bisection cut polytopes (number of variables less or equal 13) we could identify many complementarity tree, cycles with trees, and complementarity cycles with trees inequalities among the facet defining inequalities. However, so far we have figured out neither which additional conditions must be fulfilled to obtain a facet defining inequality from one of these classes nor special graph structures which guarantee supports for strong inequalities.

3.1   Complementarity Tree Inequalities

The first class of structure dependent inequalities we are going to present, bases on the tree inequality introduced in Theorems 2.2 and 2.5. We expose next the weakness of these inequalities in a general bisection case and suggest a strengthening resulting in the complementarity tree inequalities.

Theorem 3.1 (tree inequality, Conforti et al. 1990) *Let $T = (V_T, E_T)$ be a subtree of $G = (V, E)$. If $\sum_{v \in V_T} f_v > u_\tau$ then the* **tree inequality**

$$\sum_{e \in E_T} y_e \geq 1 \tag{3.1}$$

is valid for $P_{\mathcal{B}}$.

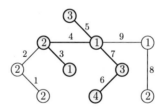

Figure 3.1: An example of a tree supporting a tree inequality $y_3 + y_4 + y_5 + y_6 + y_7 \geq 1$, $(u_\tau = 13, f(V_T) = 14)$. The numbers within nodes are the weights of nodes and the numbers at edges are edge numbers.

Proof. Let $T = (V_T, E_T)$ be a subtree of $G = (V, E)$ with $\sum_{v \in V_T} f_v > u_\tau$. Since all solutions to our problem are binary vectors, it is enough to show that an incidence vector of a bisection cut evaluating the left-hand side in (3.1) with zero does not exist. Let $\chi^{\Delta(U)}$ be an incidence vector of a cut $\Delta(U)$, $U \subseteq V$, such that

$$\sum_{e \in E_T} \chi_e^{\Delta(U)} = 0.$$

Then all nodes of T have to be in one cluster, say in U, and since

$$\sum_{v \in U} f_v \geq \sum_{v \in V_T} f_v > u_\tau$$

holds the partition $(U, V \setminus U)$ cannot be a bisection.   □

In the bisection case, particularly when the bisection ratio τ is small, to obtain a violated tree inequality one has to look for large trees with respect to the number of nodes. This pays off only during separation rounds at nodes of the branch-and-bound tree which are close to the root. In this case most variables in the solution of the linear relaxation have values close to zero. In the following extension to the tree inequality we look for the violation possibilities, when the values of the edge variables are close to 1. In the bisection case we are in the beneficial situation that we have only two clusters. Given a cut and a tree we can easily find out which nodes of the tree will be in each cluster and in this way estimate their total weight. This is how we can exploit edge variables with solution values of a relaxation close to 1.

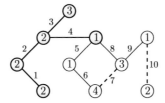

Figure 3.2: An example of a support of a complementarity tree inequality with $S = \emptyset$, $E_T = \{1, 2, 3, 4\}$, $M = \{7, 10\}$, and $u_\tau = 13$. The inequality reads $y_1 + y_2 + y_3 + y_4 + (1 - y_7) + (1 - y_{10}) \geq 1$, or simplified $y_1 + y_2 + y_3 + y_4 + 1 \geq y_7 + y_{10}$.

Theorem 3.2 (complementarity tree inequality) *Let $T = (V_T, E_T)$ be a tree in $G = (V, E)$ and let a set $M \subset E \setminus E_T$ be either empty or a matching on $V \setminus V_T$. Let a set $S \subseteq E_T$ be either empty or defined as follows. Given $C_S = \{s_1, \ldots, s_k\}$ a subset of non-adjacent nodes in V_T. For each $i \in \{1, \ldots, k\}$ let $S_i := \Delta(\{s_i\}) \cap E_T$, i.e., (V_{S_i}, S_i) is a star in T centered at s_i, and set $S := \bigcup_{i=1}^k S_i$. If*

$$\sum_{v \in V_T \setminus C_S} f_v + \sum_{vw \in M} \min\{f_v, f_w\} > u_\tau \qquad (3.2)$$

then the **complementarity tree inequality**

$$\sum_{e \in E_T \setminus S} y_e + \sum_{e \in S \cup M} (1 - y_e) \geq 1 \qquad (3.3)$$

is a valid inequality for $P_{\mathcal{B}}$.

Proof. To prove the validity of (3.3) for P_B we show that under given assumptions on the support of a complementarity tree inequality a solution $y \in P_B \cap \{0,1\}^{|E|}$ satisfying

$$\sum_{e \in E_T \setminus S} y_e + \sum_{e \in S}(1 - y_e) + \sum_{e \in M}(1 - y_e) = 0$$

does not exist. Assume on the contrary there exists a cut $\Delta(U)$, $U \subseteq V$, whose incidence vector $\chi^{\Delta(U)} \in \{0,1\}^{|E|}$ is a root of the above equality. Then

$$\chi_e^{\Delta(U)} = 0, \quad \forall e \in E_T \setminus S, \tag{3.4}$$

$$\chi_e^{\Delta(U)} = 1, \quad \forall e \in S, \tag{3.5}$$

$$\chi_e^{\Delta(U)} = 1, \quad \forall e \in M, \tag{3.6}$$

hold. By (3.4) and (3.5) all nodes in $V_T \setminus C_S$ are in one cluster, say in U. By (3.6) all edges in M are in the cut $\Delta(U)$. Thus one end-node of each edge in M belongs to V_1 and we obtain

$$\sum_{v \in U} f_v = \sum_{v \in V_T \setminus C_S} f_v + \sum_{v \in U \cap V(M)} f_v \geq \sum_{v \in V_T \setminus C_S} f_v + \sum_{vw \in M} \min\{f_v, f_w\} \geq u_\tau.$$

Thus $(U, V \setminus U)$ cannot be a bisection in G. $\qquad\square$

Remark 3.3 *Note that the assumptions in Theorem 3.2 are restricted to the bisection case and do not suffice that the complementarity tree inequalities are valid for the multiple cut polytope $P_{\mathcal{F}}$ associated with the NCGPP.*

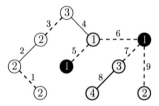

Figure 3.3: An example of a support of a complementarity tree inequality with non empty sets M and S ($u_\tau = 13$). The black labeled nodes are the centers of the stars. Thus $S = S_1 \cup S_2 = \{5\} \cup \{6,7,9\}$, $M = \{1,3\}$, and $E_T \setminus S = \{8\}$. The inequality reads $(1 - y_1) + (1 - y_3) + (1 - y_5) + (1 - y_6) + (1 - y_7) + y_8 + (1 - y_9) \geq 1$, or in a simpler form $y_1 + y_3 + y_5 + y_6 + y_7 + y_9 \leq 5 + y_8$.

Note that if S and M are empty the complementarity tree inequality becomes the generic tree inequality. In Figures 3.2 - 3.4 we show various supports of the complementarity tree inequalities depending on the existence of sets S and M.

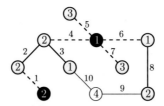

Figure 3.4: An example of a support of a complementarity tree inequality with $M = \emptyset$, $S = S_1 \cup S_2 = \{1\} \cup \{4, 5, 6, 7\}$, and $E_T \setminus S = \{2, 3, 8\}$ ($u_\tau = 13$). The inequality reads $(1-y_1)+y_2+y_3+(1-y_4)+(1-y_5)+(1-y_6)+(1-y_7)+y_8 \geq 1$, or simplified $y_1+y_4+y_5+y_6+y_7 \leq 4+y_2+y_3+y_8$.

Separation

Once we proved the validity of the complementarity tree inequalities for the polytope P_B, we turn to the separation problem (see Problem A.11) for this class of inequalities. Given a solution of a relaxation $\overline{y}$ to the MGBP, we seek for an algorithmic way

(a) either to find a complementarity tree inequality violated by $\overline{y}$, i.e.,

$$\sum_{e \in E_T \setminus S} \overline{y}_e + \sum_{e \in S \cup M} (1 - \overline{y}_e) < 1$$

for some subtree of (V_T, E_T) in G and sets $S, M \in E$ as defined in Theorem 3.2 and satisfying (3.2),

(b) or to prove that all possible complementarity tree inequalities are satisfied by $\overline{y}$.

First we comment on the complexity of the separation problem for the complementarity tree inequalities. In fact, this is a difficult task due to the quite intricate underlying graph structure and the capacity condition (3.2). So far we are able to make concrete statements for the following special cases.

(1) $S = \emptyset$ and $M = \emptyset$.

In this case our inequality reduces to the tree inequality (3.3). Assume that for some solution $\overline{y}$ of an MGBP relaxation we could solve the separation problem for the tree inequalities in polynomial time. We show that in this case we would be able to solve the separation problem of the cover knapsack inequalities (A.23). For that purpose we first bring the tree inequality into the form

$$\sum_{e \in E_T} (1 - y_e) \leq |E_T| - 1$$

which resembles the cover knapsack inequality. Consider the knapsack problem (KP) defined in Section A.5. We construct a graph G' preliminary consisting of as many nodes as items, i.e., $|U|$. Next, we add one more node w and connect it with all nodes from U. Finally we obtain a graph $G = (V, E)$ with $V = U \cup \{w\}$ and $E = \{uw : u \in U\}$. All nodes from U have the weights of the items s_u. The node w is weighted with 0.

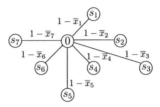

Figure 3.5: Reduction of the separation problem for cover knapsack inequalities to the separation problem for the tree inequalities.

Now, let $\bar{x}$ be a solution of a relaxation to (KP). We apply our algorithm for separation of the tree inequalities on the graph G' with respect to the solution (edge cost)

$$\tilde{y}_{uw} = 1 - \bar{x}_u.$$

and the capacity restriction $\sum_{u \in V_T} s_u > B$.

If our algorithm finds a tree inequality violated by $\tilde{y}$ then $\bar{x}$ does not satisfy the cover knapsack inequality,

$$\sum_{u \in C} x_u \leq |C| - 1,$$

with $C = V_T \setminus \{w\}$. Obviously, if our algorithm does not find any violated tree inequality in G' then all cover knapsack inequalities are satisfied with $\bar{x}$.

Since the separation problem for the knapsack cover inequalities (A.23) is NP-complete, see e.g. Ferreira [30] or Nemhauser and Vance [73], the same holds for the separation problem for the tree inequalities (3.3).

(2) $M = \emptyset$, $E_T = S$ and S consists of exactly one star.

In this case the inequality (3.3) takes the form

$$\sum_{e \in S} y_e \leq |S| - 1, \tag{3.7}$$

where (V_S, S) is a star. This inequality is similar to the cover knapsack inequality and we can also show that the separation of the cover knapsack inequalities (A.23) can be reduced to the separation problem for the inequalities (3.7). We construct a graph G as above. This time the weight of the node w does not matter as the capacity restriction for (3.7) does not include the weight of the center of the star.

Assume that we have an algorithm which solves in polynomial time the separation problem for (3.7) with respect to a solution $\bar{y}$ of a relaxation to the MGBP. We

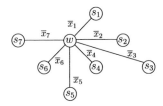

Figure 3.6: Reduction of the separation problem for cover knapsack inequalities to the separation problem for inequalities of the type (3.3).

apply this algorithm to the graph G with edge costs $\overline{y}_{uw} := \overline{x}_u$, where $\overline{x}$ is a solution of a relaxation to (KP). If we find a star (V_S, S) in G such that

$$\sum_{uw \in S} \overline{y}_{uw} > |S| - 1 \quad \text{and} \quad \sum_{uw \in S} s_u > B$$

then $\overline{x}$ violates the cover knapsack inequality

$$\sum_{u \in C} x_u \leq |C| - 1,$$

where $C = \{u \in U : uw \in S\}$.

If our algorithm does not find any violated inequality (3.7) in G then all cover knapsack inequalities are satisfied by $\overline{x}$. Thus the conclusion that separation problem for inequalities (3.7) is NP-complete follows.

Due to the NP-completeness of the separation problems for these two special cases of the complementarity tree inequality the general case is as well NP-complete. Therefore we consider a heuristic approach.

One step of our separation routine calculates maximum-weight matching. Here we apply the library *concorde* designed for solving the symmetric traveling salesman problem which is available at *http://www.tsp.gatech.edu/concorde.html*. The algorithm has a polynomial running time, $\mathcal{O}(n^3)$. Due to a quite complex design of this algorithm we refer the reader to e.g. Korte and Vygen [62] for a complete description and implementation details.

There are two conflicting criteria in searching for a violated complementarity tree inequality. The left-hand side with respect to $\overline{y}$ should be possibly close to zero or at least less than 1. That argues for a small number of edges supporting the inequality. Then again, we have to add enough nodes (and thus edges) to the support, so that the capacity of the total weight of $V_T \setminus C_S$ possibly exceeds u_τ, i.e., condition (3.2) is satisfied. Our idea is to compute a shortest path tree with respect to edge weights

$$\min\{\overline{y}_e, 1 - \overline{y}_e\}, \quad e \in E, \tag{3.8}$$

with some modifications to detect stars. We aim to achieve possibly large trees. There-fore, to stars we assign edges with values of $\overline{y}(\cdot)$ close to one. The rest of the tree is constructed with edges having solution values of a relaxation close to 0.

Our heuristic is outlined in Algorithm 3.1. For ease of exposition we denote

$$E_+ \;\; := \;\; E_T \setminus S,$$

$$V_+ \;\; := \;\; V_T \setminus C_S,$$

$$\overline{y}(E_+, S, M) \;\; := \;\; \sum_{e \in E_+} \overline{y}_e + \sum_{e \in S}(1 - \overline{y}_e) + \sum_{e \in M}(1 - \overline{y}_e).$$

We will call the subgraph of T induced by edges in E_+ the positive part of the tree. At the beginning the sets E_+, V_+, S, C_S, and M are empty (Step 1). After each update of the sets E_+, S, and M we continue only if $\overline{y}(E_+, S, M) < 1$, otherwise we stop the algorithm without success. We terminate successfully with a violated complementarity tree inequality as soon as $f(V_+) > u_r$.

We start the algorithm by selecting randomly a node $r \in V$ as the root of a tree T and add it to V_+ (Step 2). Next, we begin to compute a shortest path tree rooted at r using Dijkstra's Algorithm A.4. A candidate list L for edges which can be added to the tree is initialized with edges adjacent to r (Step 3). After each update the list is sorted with respect to weights (3.8). Let $e = vw$ be the next candidate edge with the end-node v in V_+ (Step 5). If $\overline{y}_{vw} \leq 1 - \overline{y}_{vw}$ then we update the positive part of the tree: we add e to E_+ and the node w to V_+. The cadidate list L is updated with edges adjacent to w but neither adjacent to nodes in V_+ nor in C_S (Step 7). If $\overline{y}_{vw} > 1 - \overline{y}_{vw}$, we check if we obtain a star centered at the node w (Steps 12-10). We consider the following possibilities:

(a) the edge vw is the only edge adjacent to w (see e.g. a black node with weight 2 in Figure 3.4)[1],

(b) there is at least one edge adjacent to w, say tw, with $\overline{y}_{tw} > 1 - \overline{y}_{tw}$ and $t \notin V_+$.

If (a) or (b) occurs w is marked as a center of the star, i.e., it is added to C_S. In case (a) the considered edge vw is assigned to the set of edges in stars S (Steps 10 and 13). In case (b) each edge tw such that $1 - \overline{y}_{tw} < \overline{y}_{tw}$ and $t \notin V_+$ is added to S and the corresponding end-node t to V_+ (Step 14). The candidate list L is updated accordingly with new edges adjacent to nodes recently added to V_+ (Step 15). Finally we close the loop with calculating the maximum-weight matching on $V \setminus (V_+ \cup C_S)$ with respect to the edge weights $\overline{y}$. Due to the computational effort we perform this step only after the node weight of the positive part of the tree is high enough[2], i.e., $f(V_+) \geq \frac{3}{4}u_r$. If $\overline{y}(E_+, S, M) < 1$ and

$$\sum_{v \in V_+} f_v + \sum_{vw \in M} \min\{f_v, f_w\} > u_r,$$

---

[1]We abandon the case, when $\deg(w) > 1$ and there is no edge adjacent to w, say tw, with $\overline{y}_{tw} > 1 - \overline{y}_{tw}$ and $t \notin V_+$. In this case it might pay off to add w to the positive part of the tree and gain the weight of V_+.

[2]The scalar $\frac{3}{4}$ is established empirically.

Algorithm 3.1 Separation algorithm for complementarity tree inequalities

---

Input Current solution $\overline{y}$ of a relaxation to the MGBP.

1: $success := \text{FALSE}$; $\quad E_+ := \emptyset$; $\quad V_+ := \emptyset$; $\quad S := \emptyset$; $\quad M := \emptyset$; $\quad C_S := \emptyset$;

2: randomly select a node $r \in V$; $\quad r \to V_+$;

3: initialize list $L := \{rs \in E\}$ sorted by increasing weights (3.8);

4: **repeat**

5:     take and remove the first element, say vw, from the list L, $(v \in V_+)$;

6:     **if** $\overline{y}_{vw} < 1 - \overline{y}_{vw}$ **then**

7:         $vw \to E_+$; $\quad w \to V_+$; $\quad \{ws \in E : s \notin (V_+ \cup C_S)\} \to L$;

8:     **else**

9:         **if** $\deg(w) = 1$ **then**

10:             $w \to C_S$; $\quad vw \to S$;

11:         **end if**

12:         **if** $R := \{tw : t \notin V_+, 1 - \overline{y}_{tw} < \overline{y}_{tw}\} \neq \emptyset$ **then**

13:             $w \to C_S$; $\quad vw \to S$; $\quad R \to S$;

14:             **foreach** $t : tw \in R$ **do** $t \to V_+$;

15:             **foreach** $t : tw \in R$ **do** $\{ts \in E : s \notin (V_+ \cup C_S)\} \to L$;

16:         **end if**

17:     **end if**

18:     **if** $\overline{y}(E_+, S, M) \geq 1$ **then**

19:         GO TO END;

20:     **end if**

21:     **if** $f(V_+) > u_\tau$ **then**

22:         $success := \text{TRUE}$; $\quad$ GO TO END;

23:     **end if**

24:     **if** $f(V_+) \geq \frac{3}{4} u_\tau$ **then**

25:         compute maximum-weight matching M on $V \setminus (V_+ \cup C_S)$ with rsp. to $\overline{y}(\cdot)$;

26:         **if** $\overline{y}(E_+, S, M) < 1$ **and** $f(V_+) + \sum_{vw \in M} \min\{f_v, f_w\} > u_\tau$ **then**

27:             $success := \text{TRUE}$; $\quad$ GO TO END;

28:         **else**

29:             $M := \emptyset$;

30:         **end if**

31:     **end if**

32: **until** $L = \emptyset$.

Output If *success*, support of a violated complementarity tree inequality: E_+, S, M.

---

we terminate successfully with a violated inequality, otherwise we continue computing the shortest path tree with stars (Steps 24 - 29).

3.2   Cycles with Trees Inequalities

The class of inequalities we consider next originates from cycle and cycle with tails inequalities introduced in Theorems 2.4, 2.7, and 2.8. We adapt these inequalities to the bisection case and present various versions of strengthening.

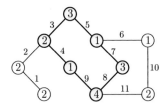

Figure 3.7: An example of a support of a cycle inequality ($u_\tau = 13$). The inequality reads $y_3 + y_4 + y_5 + y_7 + y_8 + y_9 \geq 2$.

Theorem 3.4 (cycle inequality, Conforti et al. 1990) *Let $C = (V_C, E_C)$ be a sub-cycle of $G = (V, E)$. If $\sum_{v \in V_C} f_v > u_\tau$ then the* **cycle inequality**

$$\sum_{e \in E_C} y_e \geq 2 \tag{3.9}$$

is valid for P_B.

Proof. Let χ^Δ be the incidence vector of an arbitrary bisection cut Δ in G. Since the total node weight of the cycle C exceeds u_τ, at least one edge from E_C belongs to Δ. Furthermore, each feasible cut contains an even number of edges in a cycle and thus we have $\chi^\Delta(E_C) \geq 2$. ☐

Definition 3.5 *A connected subgraph of G consisting of a cycle $C = (V_C, E_C)$ and trees $T_1 = (V_1, E_1), \ldots, T_t = (V_t, E_t)$, $t \geq 1$, such that $V_i \cap V_j \setminus V_C = \emptyset$ and $|V_i \cap V_C| = 1$ for all $1 \leq i < j \leq t$ is called* **cycle with tails** *or more precise* **cycle with t-tails**, *see Figure 3.8. The trees $T_1, \ldots, T_t$ are called* **tails**[3]. *We set $V_T := V_1 \cup \ldots \cup V_t$ and $E_T := E_1 \cup \ldots \cup E_t$.*

---

[3]In the original version of cycles with tails inequalities introduced in [31] the tails are paths. Because of the very simple extension from paths to trees we keep the name and authors of the inequality class.

A connected graph $G = (V, E)$ is called **cycles with trees decomposition** if it consists of cycles $(V_{C_1}, E_{C_1}), \ldots, (V_{C_k}, E_{C_k})$, $k \geq 1$, and trees $(V_{T_1}, E_{T_1}), \ldots, (V_{T_t}, E_{T_t})$, $t \geq 1$, such that $(V_{T_i} \cap V_{T_j}) \setminus \bigcup_{s=1}^{k} V_{C_s} = \emptyset$, $1 \leq i < j \leq t$, as well as $|V_{T_i} \cap V_{C_s}| \leq 1$ for all $s = 1, \ldots, k$ and $i = 1, \ldots, t$. Moreover, G does not contain any further cycle. To be more precise we say that in this case G has a k-**cycles with** t-**trees decomposition**. See for instance Figure 3.9.

Theorem 3.6 (cycle with tails inequality, Ferreira et al. 1996) *Let the subgraph* $G' = (V', E')$ *of* G *be a cycle with* t-*tails,* $t \geq 1$. *If* $\sum_{v \in V'} f_v > u_\tau$ *then the* **cycle with tails inequality**

$$2 \sum_{e \in E_T} y_e + \sum_{e \in E_C} y_e \geq 2$$

is valid for $P_\mathcal{B}$.

Proof. Analogous arguments as in the proofs of Theorems 3.1 and 3.4 are used. □

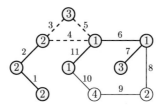

Figure 3.8: An example of a support of a cycle with 3-tails inequality: $E_C = \{3, 4, 5\}$, $E_1 = \{1, 2\}$, $E_2 = \{6, 7\}$, and $E_3 = \{11\}$ for $u_\tau = 13$. The inequality reads $2y_1 + 2y_2 + y_3 + y_4 + y_5 + 2y_6 + 2y_7 + 2y_{11} \geq 2$.

Note that the cycle with tails inequality can be seen as a strengthening of the tree inequality. We multiply both sides of the inequality by 2. We add an edge joining two nodes of the tree and get a cycle. Since each bisection cuts the cycle an even number of times, the coefficients of edges in the cycle may be reduced by 1. A cycle with tails is a 1-cycle with t-trees decomposition. If we create more cycles in the above manner, taking care that no two cycles have common edges, we obtain cycles with trees decomposition. In this sense we obtain the following strengthening of the cycle with tails inequality.

Theorem 3.7 (cycles with trees inequality) *Let* $G' = (V', E')$ *be* k-*cycles with* t-*trees in* G, $k, t \geq 1$. *Let* $E_C = \bigcup_{i=1}^{k} E_{C_i}$ *and* $E_T = \bigcup_{i=1}^{t} E_{T_i}$. *If* $\sum_{v \in V'} f_v > u_\tau$ *then the* **cycles with trees inequality**

$$2 \sum_{e \in E_T} y_e + \sum_{e \in E_C} y_e \geq 2 \tag{3.10}$$

is valid for $P_\mathcal{B}$.

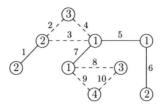

Figure 3.9: An example of a support of a 2-cycles with 3-trees inequality: $E_{C_1} = \{2,3,4\}$, $E_{C_2} = \{8,9,10\}$, $E_{T_1} = \{1\}$, $E_{T_2} = \{7\}$, and $E_{T_3} = \{5,6\}$ for $u_\tau = 13$. The inequality reads $2y_1 + y_2 + y_3 + y_4 + 2y_5 + 2y_6 + 2y_7 + y_8 + y_9 + y_{10} \geq 2$.

Proof. Analogous arguments as in the proofs of Theorem 3.1 and 3.4 are used.   □

The validity of the tree, cycle, cycle with tails, and cycles with trees inequalities bases on the idea that if we have a subgraph with node set exceeding u_τ each bisection cut will separate the support of these inequalities in at least two connected parts. In particular at least one edge from the tree (path or tail) or two edges from the cycle must be in the cut. In the strengthening presented below we give additional conditions for the supports of the cycle with tails structure so that any bisection cut will separate it into at least three connected subgraphs.

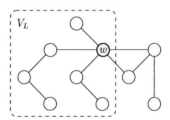

Figure 3.10: The set V_L of nodes in tails incident on node w in the support of a strengthened cycle with tails inequality.

Theorem 3.8 (strengthened cycle with tails inequality) *Let the subgraph $G' = (V', E')$ of G be a cycle with t-tails, $t \geq 1$. For some node $w \in V_C$ let $V_L := V_1 \cup \ldots \cup V_l$, $l \leq t$, be the set of nodes in tails in G' such that $\{w\} = V_1 \cap \ldots \cap V_l$, see Figure 3.10. If*

$$\sum_{v \in V_L} f_v > u_\tau \quad and \quad f_w + \sum_{v \in V' \backslash V_i} f_v > u_\tau, \quad \forall i = 1, \ldots, l, \qquad (3.11)$$

then the **strengthened cycle with tails inequality**

$$2 \sum_{e \in E_T} y_e + \sum_{e \in E_C} y_e \geq 4$$

is valid for P_B.

Proof. Note that $E_T = E(V_L) \cup E_{l+1} \cup \ldots \cup E_t$, where $E(V_L)$ denotes (as usual) set of edges connecting nodes in V_L. Let $\Delta := \Delta(U)$, $U \subseteq V$, be an arbitrary bisection cut of the graph G. Due to assumption (3.11) the total weight of V' exceeds u_τ. Hence

$$2 \sum_{e \in E_T} \chi_e^\Delta + \sum_{e \in E_C} \chi_e^\Delta \geq 2$$

is satisfied due to Theorem 3.6, i.e., $|\Delta \cap E_C| \geq 2$ or $|\Delta \cap E_T| \geq 1$. First we assume for contradiction that either $|\Delta| = |\Delta \cap E_C| = 2$ or $|\Delta| = |\Delta \cap (E_{l+1} \cup \ldots \cup E_t)| = 1$. Then V_L is completely included in one of the clusters, say in U, and by (3.11)

$$\sum_{v \in U} f_v \geq \sum_{v \in V_L} f_v > u_\tau.$$

Hence $\Delta \cap E(V_L) \neq \emptyset$ and thus

$$2 \sum_{e \in E_T} \chi_e^\Delta + \sum_{e \in E_C} \chi_e^\Delta \geq 4.$$

If $|\Delta \cap E(V_L)| > 1$ then $2 \sum_{e \in E_T} \chi_e^\Delta \geq 4$ holds trivially. Now, assume that $|\Delta| = |\Delta \cap E(V_L)| = 1$, i.e., $\Delta \cap E_i \neq \emptyset$ for some i, $1 \leq i \leq l$. Then $\{w\} \cup V' \setminus V_i$ is completely included in one of the clusters, say in U, and by (3.11) we have

$$\sum_{v \in U} f_v \geq f_w + \sum_{v \in V' \setminus V_i} f_v > u_\tau.$$

Hence $\Delta \cap (E' \setminus E_i) \neq \emptyset$ and thus also in this case

$$2 \sum_{e \in E_T} \chi_e^\Delta + \sum_{e \in E_C} \chi_e^\Delta \geq 4$$

holds. $\qquad\qquad\square$

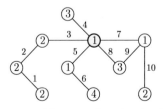

Figure 3.11: The support of the strengthened cycle with 4-tails inequality presented in Example 3.9.

Example 3.9 Consider a cycle with trees decomposition depicted in Figure 3.11, where $E_1 = \{1, 2, 3\}$, $E_2 = \{4\}$, $E_3 = \{5, 6\}$, $E_4 = \{10\}$, and $E_C = \{7, 8, 9\}$. The marked node corresponds to the node w in the assumption (3.11) and joins tails T_1, T_2 and T_3. The inequality reads

$$2y_1 + 2y_2 + 2y_3 + 2y_4 + 2y_5 + 2y_6 + y_7 + y_8 + y_9 + 2y_{10} \geq 4.$$

This inequality defines a facet of the bisection cut polytope associated with the depicted graph and the bisection bounds $l_\tau = 8$ and $u_\tau = 13$. The corresponding generic version of the cycle with tails inequality with the right-hand side equal to 2 is only valid for P_B.

Remark 3.10 *Due to the fact that the underlying supports are connected subgraphs it can be easily shown that both cycles and trees as well as strengthened cycles with tails inequalities are also valid for the cut polytope associated with NCGPP. One has to substitute u_τ with F.*

Separation

We can establish the complexity of the separation problem for cycles with trees inequalities in special cases. If the supporting graph is *0-cycles with 1-tree*, i.e., $E_C = \emptyset$ then (3.10) reduces to the tree inequality (3.1). On page 27 we have already shown that the separation problem for these inequalities is NP-complete. Another special case, that we can handle, is when $E_T = \emptyset$ and E_C consists of exactly one cycle, i.e., the underlying graph is *1-cycle with 0-trees* then (3.10) is the cycle inequality (3.9). We show that the separation problem for cycle inequalities is NP-complete by reduction from the Traveling Salesman Problem (see Problem A.18).

We are given $\overline{y} \in [0, 1]^{|E|}$, a solution of a relaxation to the MGBP. Assume we have an algorithm, which finds a violated cycle inequality, i.e., we obtain a cycle (V_C, E_C) such that

$$\sum_{e \in E_C} \overline{y}_e < 2$$

and $f(V_C) > u_\tau$. Consider now

(1) the node set V as the set of cities,

(2) $f_i = 1$ for each $i \in V$,

(3) $u_\tau = |V| - 1$, and

(4) $\overline{y}_{ij} := K d(i, j)$, where K is a constant factor scaling the distances between the cities to values in the interval $[0, 1]$.

Applying our algorithm for separation of cycle inequalities we obtain a tour of the length less than $\frac{2}{K}$. Thus we solve the Traveling Salesman Problem for $B = \frac{2}{K}$, an NP-complete problem.

Having a strong supposition that separation problem for cycles with trees inequalities is NP-complete we apply heuristic methods. They are based on separation algorithms for cycle with tails inequality introduced by Ferreira et al. in [32] with some modifications and with the strengthening presented in Theorems 3.7 and 3.8. Similarly to [32] we developed two approaches. By applying the so-called *tree based approach* we can easily strengthen the cycle with tails inequality to the cycles with tails inequality. The *cycle based approach* appears to be more convenient for the extension with the right-hand side strengthening (Theorem 3.8). Both separation heuristics are summarized in Algorithms 3.2 and 3.3.

Algorithm 3.2 Tree based separation algorithm for cycles with trees inequalities

---

Input Current solution $\overline{y}$ of a relaxation to the MGBP.

1: $success:=$FALSE ; $E_C := \emptyset$; $l(\overline{y}) := 0$;

2: compute a shortest path tree (V_T, E_T) rooted at randomly selected node s with edge weights (3.12) such that $f(V_T) > u_\tau$;

3: $l(\overline{y}) := \overline{y}(E_T)$;

4: initialize list $L := \{st \in E : s, t \in V_T, st \notin E_T\}$ sorted by increasing value (3.12) ;

5: **repeat**

6:    take and remove the first element, say st, from L ;

7:    $E_{st} = \{e \in E_T : e \text{ is on the path joining } s \text{ and } t\}$;

8:    **if** $2\overline{y}_{st} - \overline{y}(E_{st}) < 0$ and $E_{st} \cap E_C = \emptyset$ **then**

9:       $st \to E_C$; $E_{st} \to E_C$; $E_T := E_T \setminus E_{st}$; $l(\overline{y}) := l(\overline{y}) + 2\overline{y}_{st} - \overline{y}(E_{st})$;

10:   **end if**

11: **until** $L = \emptyset$

12: **if** $l(\overline{y}) < 2$ **then**

13:    $success := $ TRUE ;

14: **end if**

Output If *success*, support of a violated cycles with trees inequality E_T, E_C.

---

(1) **Tree based approach.**
At the beginning the sets E_C of edges in cycles and the set E_T of edges in trees are empty. Hence the value of the left-hand side of the inequality $l(\overline{y})$ is initialized with 0 (Step 1). In this approach we first compute a shortest path tree $T = (V_T, E_T)$ rooted at a randomly selected node $s \in V$ with respect to edge weights

$$r_{uv} := \frac{\overline{y}_{uv}}{f_v}, \quad u \in V_T, \ v \notin V_T, \tag{3.12}$$

using Dijkstra's Algorithm A.4 (Step 2). Such a rating enforces that we add to the tree edges with low costs and nodes with high weight. The computation terminates as soon as the node weight of the tree $f(V_T)$ exceeds u_r. Next, we calculate the current left-hand side value of the inequality (Step 3)

$$l(\overline{y}) := 2 \sum_{e \in E_T} \overline{y}_e.$$

If $l(\overline{y}) < 2$ we extend the tree by searching for cycles (Steps 4-11). We establish the list L of edges that are not in E_T and both their end-nodes are in V_T, sorted by increasing values $\overline{y}(\cdot)$ (Step 4). An edge $e = st$ is taken from the top of the list L (Step 6) and a path (V_{st}, E_{st}) in T joining the end-nodes s and t is computed (Step 7). We extend the support of the inequality by the edge st if the left-hand side value $l(\overline{y})$ decreases, i.e., if

$$2\overline{y}_{st} - \sum_{e \in E_{st}} \overline{y}_e < 0,$$

and if E_{st} does not have common edges with E_C (Step 8). Then st is added to E_C, edges from E_{st} are added to E_C and removed from E_T as well as $l(\overline{y})$ is accordingly updated as follows

$$l(\overline{y}) := l(\overline{y}) + 2\overline{y}_{st} - \sum_{e \in E_{st}} \overline{y}_e$$

(Step 9). The candidate st is removed from L_C and the next edge is considered. We proceed in this manner as long as the list L_C is not empty. If $l(\overline{y}) < 2$ holds and $E_C \neq \emptyset$ we end up with a violated cycles with trees inequality. If $l(\overline{y}) < 2$ holds and $E_C = \emptyset$ we obtain the tree inequality.

(2) **Cycle based approach.**
In the course of the algorithm we update the set V' with nodes of the desired cycles with trees as well as sets E_T and E_C, where we store the edges in trees and edges in cycles, respectively. All sets are initially empty (Step 1). The right-hand side value rhs is set at the beginning to 0. It takes on the value 2 or 4 depending on the strength of the inequality we finally achieve.

The first step is to find a cycle: For a randomly selected edge $ij \in E$ we compute a shortest path joining i and j with respect to the weights $\overline{y}$ using the Dijkstra's bidirectional Algorithm A.4 (Step 2). We assign the edges of the so obtained cycle to E_C and the nodes to V' (Step 3). If $y(E_C) \geq 2$ we terminate the algorithm without success, otherwise we try to expand the support with tails (Steps 4-33). For that we initialize a candidate list L with edges, whose exactly one end-node is in V' (Step 5). To add edges with possibly low costs and nodes with possibly high weight to the support of the inequality we sort edges in L by the ratio (3.12). We select edges from L and add to tails, i.e., update E_T and L accordingly, until the node set of the supporting graph V' builds a cover, i.e., $f(V') > u_r$ (Steps 6-9).

Algorithm 3.3 Cycle based separation algorithm for cycle with tails inequalities

Input Current solution $\overline{y}$ of a relaxation to the MGBP.

1: $V' := \emptyset$; $E_T := \emptyset$; $E_C := \emptyset$; $rhs := 0$;

2: for a randomly selected edge $ij \in E$ compute a shortest path (V_{ij}, E_{ij}) joining i and j with respect to edge weights $\overline{y}(\cdot)$;

3: $E_C := \{ij\} \cup E_{ij}$; $V' := V_C$;

4: **if** $\overline{y}(E_C) < 2$ **then**

5:     initialize list $L := \{tv \in E \ : \ v \in V', \ t \notin V'\}$ sorted by increasing ratio (3.12);

6:     **repeat**

7:         take and remove the first element, say tv, from list L;

8:         $tv \to E_T$; $t \to V'$; $\{st \in E \ : \ s \notin V'\} \to L$;

9:     **until** $f(V') > u_\tau$

10:     **if** $2\overline{y}(E_T) + \overline{y}(E_C) < 2$ **then**

11:         **repeat**

12:             take and remove the first element, say tv, from list L;

13:             **if** $\overline{y}_{tv} + 2\overline{y}(E_T) + \overline{y}(E_C) < 2$ **then**

14:                 $tv \to E_T$; $t \to V'$; $\{st \in E \ : \ s \notin V'\} \to L$;

15:             **end if**

16:         **until** $L = \emptyset$

17:         $rhs := 2$;   GO TO END;

18:     **else if** $2\overline{y}(E_T) + \overline{y}(E_C) < 4$ **then**

19:         **repeat**

20:             take and remove the first element, say tv, from list L;

21:             **if** $\overline{y}_{tv} + 2\overline{y}(E_T) + \overline{y}(E_C) < 4$ **then**

22:                 $tv \to E_T$; $t \to V'$; $\{st \in E \ : \ s \notin V'\} \to L$;

23:             **end if**

24:         **until** $L = \emptyset$

25:         initialize list $D := \{v \in V_C\}$ sorted by decreasing degree in $(V', E_T \cup E_C)$;

26:         **repeat**

27:             take and remove the first element, say v, from list D;

28:             **if** (3.11) holds for v **then**

29:                 $rhs = 4$;   GO TO END;

30:             **end if**

31:         **until** $D = \emptyset$ **or** $\deg(v) < 3$

32:     **end if**

33: **end if**

Output If $rhs > 0$, support of a violated cycle with tails inequality E', E_C, and rhs.

If finally

$$2 \sum_{e \in E_T} \overline{y}_e + \sum_{e \in E_C} \overline{y}_e < 2$$

then we obtain a violated cycle with tails inequality with $rhs = 2$. We add further edges from the list L to E_T as long as the left-hand side of the inequality stays below 2 (Steps 10 - 17).

If the left-hand side of the inequality exceeds 2 but not 4, we increase the right-hand side to 4 and continue extension of the tails till L is not empty. A new edge is added to E_T if the updated left-hand side stays below 4 (Steps 19 - 24). Finally we check if the so obtained strengthened cycles with trees inequality is valid for P_B. We verify condition (3.11) for nodes in the cycle ordered by their degree with respect to the support of the inequality (Step 25). The checking procedure is terminated as soon as we find a node satisfying condition (3.11), i.e., we have a valid inequality, or the degree of the selected node is less than 3, i.e., the remaining nodes are not adjacent to tails. In the latter case our procedure terminates without a success (Steps 26 - 31).

Both algorithms were partially developed by Jetchev [51].

3.3   Complementarity Cycles with Trees Inequalities

The drawback of searching for violated cycles with trees inequalities is the same as in the case of tree inequalities. They can be found only in the nodes of the branch-and-bounfd tree which are close to the root. Hence we search for conditions, when the complementing of variables is possible. The new class of inequalities presented in this section is based on the following idea. We consider a subgraph cycles with trees G' and a subset D of edges in G'. Assume that D is a feasible cut in G', in the sense that each cycle in G' is cut an even number of times. The cut D indicates a bipartition in G', say (V_1', V_2'). If the node weight of one of the clusters, V_1' or V_2', exceeds u_τ, D cannot be a bisection cut. Should the node weight of both clusters be less than u_τ, we can extend the bigger one, say V_1', by considering the nodes in the remaining part of the graph G, i.e., in $V \setminus (V_1' \cup V_2') =: \overline{V}$. For that we compute a matching M on $E(\overline{V})$ and consider the cut $\Delta := D \cup M$. Now, we add the "lighter" end-nodes of edges in M to V_1'. This node set corresponds to one of the clusters from the bipartition due to Δ. If its weight exceeds u_τ, the cut Δ cannot be a bisection cut. See also Example 3.12. To eliminate such cuts like Δ is the task of complementarity cycles with trees inequality formalized below.

Theorem 3.11 (complementarity cycles with trees inequality) *Let a subgraph $G' = (E', V')$ in G be a k-cycles with t-trees, $k, t \geq 1$, with $E_C = \bigcup_{i=1}^{k} E_{C_i}$ and*

$E_T = \bigcup_{i=1}^{t} E_{T_i}$. Let $\emptyset \subseteq D$ be a cut in G', i.e., a subset of E' such that $D \cap E_{C_i} = \emptyset$ or $|D \cap E_{C_i}|$ is even for each $i = 1, \ldots, k$. Let M be a matching on $E(V \setminus V')$. Furthermore, let (V_1', V_2') be the bipartition in G' corresponding to the cut D. If

$$\max\{f(V_1'), f(V_2')\} + \sum_{vw \in M} \min\{f_v, f_w\} > u_\tau \tag{3.13}$$

then the **complementarity cycles with trees inequality**

$$\sum_{e \in (D \cap E_T) \cup M} 2(1 - y_e) + \sum_{e \in E_T \setminus D} 2y_e + \sum_{e \in D \cap E_C} (1 - y_e) + \sum_{e \in E_C \setminus D} y_e \geq 2 \tag{3.14}$$

is a valid inequality for $P_\mathcal{B}$.

Proof. We only need to show that a bisection cut whose incidence vector yields a value less than 2 on the left-hand side of the inequality (3.14) does not exist. Assume there is a cut $\Delta := \Delta(U)$, $U \subseteq V$, such that

$$\sum_{e \in (D \cap E_T) \cup M} 2(1 - \chi_e^\Delta) + \sum_{e \in E_T \setminus D} 2\chi_e^\Delta + \sum_{e \in D \cap E_C} (1 - \chi_e^\Delta) + \sum_{e \in E_C \setminus D} \chi_e^\Delta = 0.$$

Since χ^Δ is a binary vector, we obtain

$$\chi_e^\Delta \;=\; 1, \qquad \forall\, e \in D \text{ and } \forall\, e \in M,$$

$$\chi_e^\Delta \;=\; 0, \qquad \forall\, e \in E' \setminus D,$$

and hence Δ coincides with D in E'. W.l.o.g., we assume that $V_1' \subseteq U$ and $V_2' \subseteq V \setminus U$, as well as $f(V_1') \geq f(V_2')$. By assumption (3.13) we have

$$f(U) \geq f(V_1') + \sum_{vw \in M} \min\{f_v, f_w\} > u_\tau$$

and thus $(U, V \setminus U)$ cannot be a bisection in G.

Now, assume that

$$\sum_{e \in (D \cap E_T) \cup M} 2(1 - \chi_e^\Delta) + \sum_{e \in E_T \setminus D} 2\chi_e^\Delta + \sum_{e \in D \cap E_C} (1 - \chi_e^\Delta) + \sum_{e \in E_C \setminus D} \chi_e^\Delta = 1.$$

This is only possible if

$$\sum_{e \in D \cap E_C} (1 - \chi_e^\Delta) + \sum_{e \in E_C \setminus D} \chi_e^\Delta = 1,$$

i.e.,

$$-\sum_{e \in D \cap E_C} \chi_e^\Delta + \sum_{e \in E_C \setminus D} \chi_e^\Delta = 1 - |D \cap E_C|.$$

Due to our assumption $|D \cap E_C|$ is an even number. Hence the left-hand side of the last equality must be an odd number. This means that Δ contains an odd number of edges from some cycle in G' and thus it cannot be a cut in G. $\qquad\square$

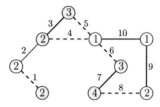

Figure 3.12: The support of the complementarity cycles with trees inequality considered in Example 3.12.

Example 3.12 Consider the graph plotted in Figure 3.12 and the support of the complementarity cycles with trees inequality with $E_M = \{1\}$, $D = \{4, 5, 6, 8\}$, $E_C = \{3, 4, 5\} \cup \{6, 7, 8, 9, 10\}$, and $E_T = \emptyset$. The cut D (dashed edges) partitions the nodes of the graph into the "white" and the "gray" cluster. The node weight of the "gray" cluster amounts to 14 and exceeds $u_\tau = 13$. The inequality reads

$$2(1 - y_1) + y_3 + (1 - y_4) + (1 - y_5) + (1 - y_6) + y_7 + (1 - y_8) + y_9 + y_{10} \geq 2,$$

or in simplified form

$$4 + y_3 + y_7 + y_9 + y_{10} \geq y_1 + y_4 + y_5 + y_6 + y_8,$$

and defines a facet of the bisection cut polytope associated with the depicted graph for bisection bounds $l_\tau = 7$ and $u_\tau = 13$.

Separation

The complementarity cycles with trees inequalities can be seen as a strengthening of the complementarity tree and cycles with trees inequality. Hence the complexity of the separation problem can be deduced from the considerations on the complexity of the separation problems for these inequalities presented in previous sections.

In our heuristic, outlined in Algorithm 3.4, we use the fact that given a cut and a subtree in G it is easy to calculate the total weight of nodes of the tree assigned to each of the two clusters with respect to this cut.

For ease of exposition we divide the set of edges E of the graph G into two sets

$$\begin{aligned} E_- &:= \{e \in E : 1 - \overline{y}_e < \overline{y}_e\}, \\ E_+ &:= E \setminus E_-. \end{aligned}$$

Since the sets M and E_C may not be updated at all during the algorithm, we initialize them as empty sets (Step 1).

The first milestone is to compute a shortest path tree $T = (V_T, E_T)$ with respect to edge weights

$$\omega_e := \min\{\overline{y}_e, 1 - \overline{y}_e\}, \quad \forall e \in E, \tag{3.15}$$

using Dijkstra's Algorithm A.4 (Step 2). To obtain a tree with possibly small weight $\omega(E_T)$, and thus minimize the left-hand side of the inequality, we add an edge e to the tree only if $\omega_e < 0.1$.

Next, we calculate the bipartition (V_1', V_2') in T with respect to the cut $E_T \cap E_- =: D$ (Step 3) and initialize the cluster incidence vector $c \in \{0, 1\}^{|V_T|}$ as follows

$$c := \begin{cases} 1, & v \in V_1', \\ 0, & v \in V_2', \end{cases}$$

(Step 4). Only if none of the clusters V_1', V_2' defines a cover (Step 6), we compute the minimum-weight matching[4] M on the subgraph induced by $V \setminus V_T$ with respect to edge weights $1 - \overline{y}(\cdot)$ (Step 7). This is due to the fact that the computing of the matching is rather costly. To keep the left-hand side of the inequality as small as possible, we select edges from the matching M with the smallest weights and add them to the support (the set D) of the inequality until the condition (3.13) is satisfied (Steps 8 - 11). Only then the tree is upgraded to cycles with trees (Steps 14 - 23). All edges in $E \setminus E_T$ having both end-nodes in V_T are sorted by increasing weights (3.15).

A new edge $e = st$ together with edges on the path (V_{st}, E_{st}) joining nodes s and t in T are added to E_C (Step 19 and 21) provided the following three conditions are satisfied (*if-statement* in Steps 18 and 20):

(1) either $\overline{y}_{st} < 1 - \overline{y}_{st}$ $(st \in E_+)$ and s, t are in one cluster of the bipartition (V_1', V_2'), or $1 - \overline{y}_{st} < \overline{y}_{st}$ $(st \in E_-)$ and s, t are in different clusters,

(2) none of the edges in E_{st} is in E_C,

(3) by adding the edge st to the support of the inequality the current value on the left-hand side with respect to $\overline{y}$ is reduced.

If conditions (1) - (3) are satisfied for an edge $st \in E_-$ (*if-statement* in Step 20) then we update also the set D with the edge st.

Finally, we remove from E_T all common edges with E_C and verify if the cycles with trees inequality supported by the sets D, E_T and E_C is violated by the solution $\overline{y}$, i.e., if

$$lhs := \sum_{e \in D \setminus E_C} 2(1 - \overline{y}_e) + \sum_{e \in E_T \setminus D} 2\overline{y}_e + \sum_{e \in D \cap E_C} (1 - \overline{y}_e) + \sum_{e \in E_C \setminus D} \overline{y}_e < 2$$

(Step 26). Note that we earlier merged some edges from the matching M with the set D. Once we obtained a violated inequality we can still strengthen it by adding the remaining edges from the matching M as long as the updated left-hand side does not exceed the value 2 (Steps 28 - 30).

---

[4]Here we use also the *concorde* library described in Section 3.1.

Algorithm 3.4 Separation algorithm for complementarity cycles with trees inequalities

Input Current solution $\overline{y}$ of a relaxation to the MGBP.

1: $E_- := \{e \in E \;:\; 1 - \overline{y}_e < \overline{y}_e\}$; $E_+ := E \setminus E_-$; $E_C := \emptyset$; $M := \emptyset$;
    $success = \text{FALSE}$;

2: randomly select node $s \in V$ and compute a shortest path tree (V_T, E_T) rooted at s
    with edge weights (3.15) not exceeding 0.1;

3: calculate bipartition (V_1', V_2') in T with respect to the cut $D := E_T \cap E_-$;

4: **foreach** $v \in V_T$ **do** $c[v] = 1$ for $v \in V_1'$; **and** $c[v] = 0$ for $v \in V_2'$;

5: $f_{\max} := \max\{f(V_1'), f(V_2')\}$;

6: **if** $f_{\max} \leq u_\tau$ **then**

7:    compute minimum-weight matching M on $E(V \setminus V_T)$ with edge weights $1 - \overline{y}(\cdot)$;

8:    sort edges in M by increasing weights $1 - \overline{y}(\cdot)$;

9:    **repeat**

10:       $M \to uv$; $uv \to D$; $f_{\max} := f_{\max} + \min\{f_u, f_v\}$;

11:    **until** $M = \emptyset$ **or** $f_{\max} > u_\tau$

12: **end if**

13: **if** $f_{\max} > u_\tau$ **then**

14:    initialize list $L := \{st \in E \setminus E_T \;:\; s, t \in V_T\}$ sorted by increasing weights (3.15);

15:    **repeat**

16:       take and remove the first element, say st, from list L;

17:       $E_{st} := \{e \in E_T \;:\; e$ is on the path joining s and $t\}$;

18:       **if** $st \in E_+$ **and** $c[s] = c[t]$ **and** $E_{st} \cap E_C = \emptyset$ **and** $\overline{y}_{st} < \omega(E_{st})$ **then**

19:          $st \to E_C$; $E_{st} \to E_C$;

20:       **else if** $st \in E_-$ **and** $c[s] := c[t]$ **and** $E_{st} \cap E_C = \emptyset$ **and** $1 - \overline{y}_{st} < \omega(E_{st})$ **then**

21:          $st \to D$; $st \to E_C$; $E_{st} \to E_C$;

22:       **end if**

23:    **until** $L = \emptyset$

24:    $E_T := E_T \setminus E_C$;

25:    $lhs := \sum_{e \in D \setminus E_C} 2(1 - \overline{y}_e) + \sum_{e \in E_T \setminus D} 2\overline{y}_e + \sum_{e \in D \cap E_C}(1 - \overline{y}_e) + \sum_{e \in E_C \setminus D} \overline{y}_e$;

26:    **if** $lhs < 2$ **then**

27:       $success = \text{TRUE}$;

28:       **while** $M \neq \emptyset$ **do**

29:          $M \to e$;

30:          **if** $lhs + 2(1 - \overline{y}_e) < 2$ **then** $e \to D$; **else** $M := \emptyset$;

31:       **end while**

32:    **end if**

33: **end if**

Output If $success$, support of a violated complementarity cycles with trees inequality
D, E_T, and E_C.

3.4   Odd Cycle Inequalities

For the sake of completeness we cite in this section the validity proof for odd cycle inequalities for P_B as well as the polynomial time separation algorithm due to Barahona and Mahjoub [11]. Unlike their name the inequalities rely on the fact that each cut contains an even number of edges from any cycle in G.

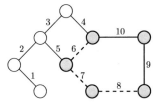

Figure 3.13: An example of a support of an odd cycle inequality with $E_C = \{6, 7, 8, 9, 10\}$ and $D = \{6, 7, 8\}$ (dashed edges). The inequality reads $-y_6 - y_7 - y_8 + y_9 + y_{10} \leq 2$.

Theorem 3.13 (Barahona, Mahjoub 1986) *Let $C = (E_C, V_C)$ be a cycle in G. Let D be a subset of E_C such that $|D|$ is odd. Then the* **odd cycle inequality**

$$\sum_{e \in D} y_e - \sum_{e \in E_C \setminus D} y_e \leq |D| - 1 \tag{3.16}$$

is a valid inequality for the cut polytope P_C.

Proof. We rewrite (3.16) in the following form

$$\sum_{e \in D} (1 - y_e) + \sum_{e \in C \setminus D} y_e \geq 1, \tag{3.17}$$

for a given cycle $C = (V_C, E_C)$ and a set $D \subseteq E_C$ of an odd cardinality. Let $\overline{y} \in \{0, 1\}^{|E|}$ be some vector such that

$$\sum_{e \in D} (1 - \overline{y}_e) + \sum_{e \in C \setminus D} \overline{y}_e = 0.$$

The above equality is satisfied if and only if

$$\overline{y}_e = 1, \quad \forall e \in D,$$
$$\overline{y}_e = 0, \quad \forall e \in E_C \setminus D.$$

This means that $\overline{y}(E_C)$ is an odd number and thus $\overline{y}$ cannot be an incident vector of a cut in G. $\qquad \square$

Remark 3.14 *(3.16) is obviously valid for $P_{\mathcal{B}}$ due to the fact that $P_{\mathcal{B}} \subseteq P_C$. Barahona and Mahjoub showed in [11] that the odd cycle inequalities are facet defining for P_C if and only if C is a chordless cycle, i.e., there is no edge $uv \in E$ such that $u, v \in V_C$ and $uv \notin E_C$. Since $P_{\mathcal{B}} \subseteq P_C$, we obtain that this condition is at least necessary for (3.16) to define a facet of $P_{\mathcal{B}}$.*

Separation

Barahona and Mahjoub [11] give a polynomial time algorithm which solves the separation problem for odd cycle inequalities. We cite briefly the idea of this algorithm and put together the main steps in Algorithm 3.5.

Algorithm 3.5 Separation Algorithm for Odd Cycle Inequalities

---

Input Current solution $\overline{y}$ of a relaxation to the MGBP.

1: $i := 0$; (counter for supports violating the odd cycle inequality)

2: create auxiliary graph $G' = (V', E')$:

3:    for each $v \in V$ add node v' ;

4:    for each $vu \in E$ add edges vu', $v'u$, and $v'u'$ (see Figure 3.14) ;

5:    set edge weights $\omega_{vu} = \omega_{v'u'} = \overline{y}_{vu}$, $\omega_{v'u} = \omega_{vu'} = 1 - \overline{y}_{vu}$;

6: **for each** $v \in V$ **do**

7:    compute shortest path $(V_{vv'}, E_{vv'})$ joining v in v' with respect to ω ;

8:    check if $(V_{vv'}, E_{vv'})$ induces a cycle in the original graph G ;

9:    **if** $\omega(E_{vv'}) < 1$ **then**

10:      $i \rightarrow i + 1$;

11:      **for each** $e \in E_{vv'}$ **do**

12:        **if** $e = ut'$ **or** $e = u't$, $(u, t \in V, u', t' \in V' \setminus V)$ **then**

13:          $ut \rightarrow D_i$;

14:        **else**

15:          $ut \rightarrow \overline{D}_i$;

16:        **end if**

17:      **end for**

18:    **end if**

19: **end for**

Output If $i > 0$, list of supports $(D_1, \overline{D}_1) \ldots (D_i, \overline{D}_i)$ such that $|D_j|$ is odd and $\overline{y}(D_j) - \overline{y}(\overline{D}_j) > |D_j| - 1$ for each j, $1 \leq j \leq i$.

Consider the odd cycle inequality in the form (3.17). Given $\overline{y}$, a solution of a relaxation to the MGBP, we look for a minimum weighted cycle, where some edges have the weight $\overline{y}(\cdot)$ and an odd number of edges has the weight $1 - \overline{y}(\cdot)$. To this end, an auxiliary graph is built from G (Steps 2-5 and Figure 3.14). Then a shortest path is computed from each original node to its copy. We apply Dijkstra's Algorithm A.4, as usual. Once we found such a path we check if it induces a cycle in the original graph G (Step 8). It can happen that we obtain for instance two cycles connected by one node. We take that cycle with the smaller edge weight with respect to ω containing an appropriate odd subset of edges, see Steps 12-13. The edge weight of the path is the weight of the required cycle (Step 9). If it is less than 1 we obtain a support of a violated odd cycle inequality (Steps 11-15). To achieve the copy destination node we have to pass an odd number of edges joining original and copied nodes. These edges define the set D of an odd cardinality.

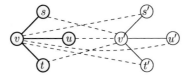

Figure 3.14: An example of an auxiliary graph considered in the separation algorithm for odd cycle inequalities.

The algorithm has the complexity $\mathcal{O}(|V|^3)$ due to the fact that for each node we calculate a shortest path with the complexity $\mathcal{O}(|V|^2)$.

Chapter 4

Knapsack Tree Inequalities

Knapsack tree inequalities were first introduced by Ferreira, Martin, de Souza, Weismantel, and Wolsey in [31] as valid inequalities for the polytope $P_{\mathcal{F}}$ associated with the node capacitated graph partitioning problem (NCGPP), introduced in Section 2.2. They build a rare class of inequalities which exploit the knapsack condition in the problem specification. Node weights are coupled with edge variables using a tree spanned on these nodes. This tree and the knapsack condition on the node weights give the name to the inequality class.

Ferreira et al. [32] empirically showed the strength of the knapsack tree inequalities. In this chapter we justify theoretically this computational success with respect to the bisection case. We present several strengthening methods for these inequalities as well as necessary and sufficient conditions, when they define facets of the bisection cut polytope.

The knapsack tree inequalities are defined as follows. Let

$$\sum_{v \in V} a_v x_v \leq a_0,$$

with $a_v \geq 0$ for all $v \in V$, be a valid inequality for the knapsack polytope $P_{\mathcal{K}}$, defined in Section 2.2. Select a node $r \in V$ and a subtree $T = (V_T, E_T)$ of G rooted at r. For $v \in V_T$ let $\Pi_{rv} = (V_{rv}, E_{rv})$ be the path in T joining v with root r. The inequality

$$\sum_{v \in V_T} a_v \left(1 - \sum_{e \in E_{rv}} y_e\right) \leq a_0 \tag{4.1}$$

is called **knapsack tree inequality**. The underlying tree T is called **knapsack tree**.

Proposition 4.1 *The knapsack tree inequality (4.1) is valid for the polytope $P_{\mathcal{B}}$.*

Proof. Let $(S, V \setminus S)$ be a bisection in G and, w.l.o.g., assume that $r \in S$. Define $U \subseteq S \cap V_T$ as follows

$$U := \{ s \in V_T \; : \; V_{rs} \subseteq S \},$$

49

i.e., U contains all nodes whose paths to root node r have all nodes in S. Then $\sum_{e \in E_{rv}} y_e = 0$ for all $v \in U$ and $\sum_{e \in E_{rv}} y_e \geq 1$ for all $v \notin U$. Using the fact that $a_v \geq 0$ for all $v \in V$ we obtain

$$\sum_{v \in V_T} a_v \left(1 - \sum_{e \in E_{rv}} y_e\right) \leq \sum_{v \in U} a_v + \sum_{v \in V_T \setminus U} a_v \left(1 - \sum_{e \in E_{rv}} y_e\right) \leq \sum_{v \in S} a_v \leq a_0.$$

See also Proposition 3.11 in [31]. $\square$

In the sequel we denote by V_e^r be the set of nodes, whose path to r contains $e \in E_T$, see Figure 4.1.

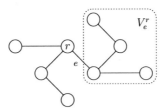

Figure 4.1: An example of the node set V_e^r.

4.1 Strengthened Knapsack Tree Inequalities

The general form (4.1) of the knapsack tree inequality can be strengthened in several ways as we show in the following section. We start with the standard method which can be applied to any valid inequality for a binary polytope. The next two strengthening ways exploit the structure of the underlying knapsack tree. The first one uses the length of the paths connecting nodes with the root node with respect to the number of edges. The second one concerns the weight of the nodes on these paths.

4.1.1 0-1 Programming Strengthening

If we determine explicitly the coefficients of inequality (4.1) then it takes the form

$$\sum_{e \in E_T} \left(\sum_{v \in V_e^r} a_v\right) y_e \geq \sum_{v \in V_T} a_v - a_0.$$

Applying the strengthening given in Proposition A.7 to the above inequality we obtain

$$\sum_{e \in E_T} \min\left\{\sum_{v \in V_e^r} a_v, \sum_{v \in V_T} a_v - a_0\right\} y_e \geq \sum_{v \in V_T} a_v - a_0. \tag{4.2}$$

To emphasize that the coefficients in (4.2) depend on root node r we introduce the notation

$$\alpha_0 := \sum_{v \in V_T} a_v - a_0, \qquad \alpha_e^r := \min\{\sum_{v \in V_e^r} a_v, \alpha_0\}, \, e \in E_T, \tag{4.3}$$

and consider (4.2) in the form

$$\sum_{e \in E_T} \alpha_e^r y_e \geq \alpha_0, \tag{4.4}$$

or $(\alpha^r)^T y \geq \alpha_0$ for short.

Note that if we change the root of T the right-hand side of (4.4) remains the same, since by this operation we do not eliminate nodes in V_T.

Proposition 4.2 *The strengthened knapsack tree inequality (4.4) is valid for the polytope $P_{\mathcal{B}}$.*

Proof. It follows from Proposition A.7. $\qquad \qquad \square$

Remark 4.3 *In [31] (Proposition 3.12) the following strengthening of inequality (4.1)*

$$\sum_{e \in E_T} \min\{\sum_{v \in V_e^r} a_v, \alpha_0, a_0\} y_e \geq \sum_{v \in V_T} a_v - a_0$$

is presented and proven to be valid for $P_{\mathcal{F}}$. In the bisection case the above inequality reduces to (4.2). This follows from the fact that $\alpha_0 \leq a_0$, as explained below.
In the bisection case if $\bar{x} = (\bar{x}_1, \ldots, \bar{x}_n)^T \in P_{\mathcal{K}} \cap \{0, 1\}^n$ then also $\tilde{x} = (1 - \bar{x}_1, \ldots, 1 - \bar{x}_n)^T$ lies in $P_{\mathcal{K}} \cap \{0, 1\}^n$. This follows from the fact that the total node weight of each of the clusters $\{v \in V : \bar{x}_v = 1\}$ and $\{v \in V : \bar{x}_v = 0\}$ cannot exceed u_r. Given $\sum_{v \in V} a_v x_v \leq a_0$, a valid inequality for the knapsack polytope $P_{\mathcal{K}}$, if

$$\sum_{i \in V} a_v \bar{x}_v \leq a_0$$

then also

$$\sum_{v \in V} a_v \tilde{x}_v = \sum_{v \in V} a_v (1 - \bar{x}_v) \leq a_0.$$

Summing up these two inequalities yields $\sum_{v \in V} a_v \leq 2a_0$ and thus

$$\alpha_0 = \sum_{v \in V} a_v - a_0 \leq a_0.$$

4.1.2   Shortest Path Tree Strengthening

It turns out that the strength of a knapsack tree inequality depends on the underlying knapsack tree. A stronger inequality is obtained if a sub-path of a path to root can be replaced by a single edge from the original graph.

Proposition 4.4 *Let $\sum_{v \in V} a_v x_v \leq a_0$, with $a_v \geq 0$ for all $v \in V$, be a valid inequality for $P_{\mathcal{K}}$. Let T and T' be knapsack trees, both rooted at one node r, and such that $V_T = V'_T$. Let $\Pi_{rv} = (V_{rv}, E_{rv})$, $\Pi'_{rv} = (V'_{rv}, E'_{rv})$, $v \in V_T$ be the paths joining nodes v and r in T and T' respectively. If $V'_{rv} \subseteq V_{rv}$ for each node v then*

$$\sum_{e \in E_T} (\sum_{v \in V_e^r} a_v) y_e \geq \sum_{e \in E_{T'}} (\sum_{v \in V_e^r} a_v) y_e \geq \sum_{v \in V_T} a_v - a_0 \qquad (4.5)$$

for all $y \in P_{\mathcal{B}}$.

Proof. Let T and T' be as defined above. Furthermore, we assume that $V'_{rv} \subseteq V_{rv}$ holds for each node $v \in V_T (= V_{T'})$. First we are going to show that

$$\sum_{v \in V_T} a_v (1 - \sum_{e \in E_{rv}} y_e) \leq \sum_{v \in V_T} a_v (1 - \sum_{e \in E'_{rv}} y_e) \qquad (4.6)$$

holds and then derive (4.5).

For all $v \in V_T$ such that $V'_{rv} = V_{rv}$ we have $E'_{rv} = E_{rv}$ and

$$a_v (1 - \sum_{e \in E_{rv}} y_e) = a_v (1 - \sum_{e \in E'_{rv}} y_e). \qquad (4.7)$$

Now, consider $v \in V_T$ such that $V'_{rv} \subset V_{rv}$. In this case the graph induced by $E_{rv} \cup E'_{rv}$ contains a set of cycles $(V_{C_1}, E_{C_1}), \dots, (V_{C_l}, E_{C_l})$, $l \geq 1$. W.l.o.g., we assume that (V_{C_1}, E_{C_1}) is the closest cycle to r with respect to number of nodes on the path joining r and some node in V_{C_1}. Because of $V'_{rv} \subset V_{rv}$ we have $|E_{C_1} \cap E_{T'}| = 1$. Denote this

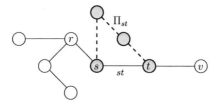

Figure 4.2: Replacing path Π_{st} by edge st in cycle (V_{C_1}, E_{C_1}).

edge by st, see Figure 4.2. Note that $E_{rs} = E'_{rs}$ and $\Pi_{st} = (V_{st}, E_{st})$ is the sub-path of Π_{rv} with $E_{st} \cap E'_{rv} = \emptyset$. Hence st and Π_{st} define cycle (V_{C_1}, E_{C_1}). Thus

$$\sum_{e \in E_{st}} y_e - y_{st} \geq 0$$

is a valid odd cycle inequality for P_B (see Section 3.4). Applying this inequality we obtain

$$\sum_{e \in E_{rt}} y_e = \sum_{e \in E_{rs}} y_e + \sum_{e \in E_{st}} y_e \geq \sum_{e \in E'_{rs}} y_e + y_{st} = \sum_{e \in E'_{rt}} y_e.$$

Similarly we deal with the cycles $C_2, \ldots, C_l$, if $l \geq 2$, and obtain

$$\sum_{e \in E_{rv}} y_e \geq \sum_{e \in E'_{rv}} y_e.$$

Hence

$$a_v(1 - \sum_{e \in E_{rv}} y_e) \leq a_v(1 - \sum_{e \in E'_{rv}} y_e).$$

holds for any node $v \in V_T$ and thus also (4.6). It remains to show (4.5). By (4.6) and using the assumption that $V_T = V_{T'}$ we obtain

$$\sum_{e \in E_T} (\sum_{v \in V_e^r} a_v) y_e = \sum_{v \in V_T} a_v - \sum_{v \in V_T} a_v (1 - \sum_{e \in E_{rv}} y_e) \geq \sum_{v \in V_T} a_v - \sum_{v \in V_T} a_v (1 - \sum_{e \in E'_{rv}} y_e)$$

$$= \sum_{e \in E_{T'}} (\sum_{v \in V_e^r} a_v) y_e \geq \sum_{v \in V_{T'}} a_v - a_0 = \sum_{v \in V_T} a_v - a_0.$$

$\square$

Having the above result one can ask, if demanding that the knapsack tree is a shortest path tree spanned on V_T yields the strongest inequality in the sense given in Proposition 4.4. We use an example to show that the answer is negative. Let knapsack trees T and T' be spanned on a node set V_T and rooted at a node r. Let $\Pi_{rv} = (V_{rv}, E_{rv})$, $\Pi'_{rv} = (V'_{rv}, E'_{rv})$, $v \in V_T$, be the paths joining nodes v and r in T and T' respectively and

$$\sum_{v \in V_T} a_v (1 - \sum_{e \in E_{rv}} y_e) \leq a_0, \tag{4.8}$$

$$\sum_{v \in V_T} a_v (1 - \sum_{e \in E'_{rv}} y_e) \leq a_0 \tag{4.9}$$

be the corresponding knapsack tree inequalities. Assume that T' is a shorter path tree than T and there exists exactly one node $s \in V_T$ such that $|E_{rs}| > |E'_{rs}|$. Since Π'_{rs} does not coincide with Π_{rs} and both paths join s with root r, there exists a cycle $C = (V_C, E_C)$ such that

$$E_C = (E'_{rs} \setminus E_{rs}) \cup (E_{rs} \setminus E'_{rs}),$$

see Figure 4.3. Furthermore, $E_{rs} \setminus E_C = E'_{rs} \setminus E_C =: \bar{E}$, i.e., both paths differ only in

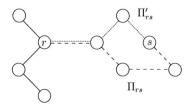

Figure 4.3: Replacing path Π_{rs} by path Π'_{rs}.

the edges from the cycle. The terms corresponding to node s in inequalities (4.8) and (4.9) take the form

$$a_s\Big(1 - \sum_{e \in E_{rs}} y_e\Big) = a_s\Big(1 - \sum_{e \in \bar{E}} y_e - \sum_{e \in E_{rs} \cap E_C} y_e\Big),$$

$$a_s\Big(1 - \sum_{e \in E'_{rs}} y_e\Big) = a_s\Big(1 - \sum_{e \in \bar{E}} y_e - \sum_{e \in E'_{rs} \cap E_C} y_e\Big),$$

respectively. Then the inequality

$$a_s\Big(1 - \sum_{e \in E_{rs}} y_e\Big) \le a_s\Big(1 - \sum_{e \in E'_{rs}} y_e\Big)$$

holds if

$$\sum_{e \in E_{rs} \cap E_C} y_e \ge \sum_{e \in E'_{rs} \cap E_C} y_e. \qquad (4.10)$$

Hence the validity of (4.10) yields a necessary condition for (4.9) to be a strengthening of (4.8). Although (4.10) is valid for P_B only if $|E'_{rs} \cap E_C| = 1$ (this fact we used in the proof of Proposition 4.4) the condition can still be useful in the separation algorithm for the knapsack tree inequalities as we show in Section 4.3.

4.1.3   Minimum Root Strengthening

Assume we have established a tree $T = (V_T, E_T)$ rooted at r which yields the strongest knapsack tree inequality in the sense (4.5). We can further reduce the coefficients by moving root r along the tree. If we replace r by another node from V_T the paths change and thus the values of the coefficients in the corresponding knapsack tree inequality. We are going to show that the strongest or in some cases even a facet-defining inequality is achieved if r corresponds to a sort of equilibrium with respect to the cumulated node weights on the paths to r. Since further improvements pay off only if a stronger inequality than (4.4) is achieved, we act on knapsack tree inequalities in this form.

At first we derive some relations based on the definition of the coefficients α_e^r, $r \in V_T$, $e \in E_T$ which we exploit in the proofs of the results presented in this section. In the next lemma we investigate the change of coefficients if the root is moved from the node r to an adjacent node s.

Lemma 4.5 *Let* $T = (V_T, E_T)$ *be a tree in* G. *Furthermore, let* $r, s \in V_T$ *be two adjacent nodes with* $\bar{e} = rs \in E_T$. *Then we have*

(1) $\alpha_e^r = \alpha_e^s$ *for all* $\bar{e} \neq e$, $e \in E_T$

(2) $\sum_{e \in E_T} \alpha_e^r < \sum_{e \in E_T} \alpha_e^s$ *if and only if* $\alpha_{\bar{e}}^r < \alpha_{\bar{e}}^s$ *and*

$\sum_{e \in E_T} \alpha_e^r = \sum_{e \in E_T} \alpha_e^s$ *if and only if* $\alpha_{\bar{e}}^r = \alpha_{\bar{e}}^s$.

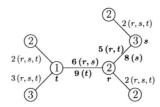

Figure 4.4: The α-coefficients with respect to roots r, s, and t. The number corresponds to value of the coefficient with respect to the roots in the brackets behind it, e.g. $6(r,s)$ at edge tr means that $\alpha_{tr}^r = \alpha_{tr}^s = 6$. One can see that the coefficients differ at edges tr and rs joining the roots.

Proof. (1) For $e \neq rs$ we have $\{v : e \in P_{rv}\} = \{v : e \in P_{sv}\}$ and thus

$$\sum_{v:e\in P_{rv}} a_v = \sum_{v:e\in P_{sv}} a_v,$$

see also Figure 4.4.

(2) The claim follows directly from (1). $\qquad\square$

Lemma 4.6 *Let* T *be a tree in* G *rooted at node* r *and let* e *precede* f *on the path to* r. *Then*

$$\alpha_f^r \leq \alpha_e^r, \tag{4.11}$$

i.e., the closer an edge is to root r *with respect to the number of edges, the bigger is its coefficient.*

Proof. W.l.o.g., we assume that e and f are adjacent. Setting $e := ij$ and $f := jk$ we obtain

$$\sum_{v:e\in P_{rv}} a_v = \sum_{v:f\in P_{rv}} a_v + a_j + \sum_{\bar{e}\in\bar{E}} \left(\sum_{v:\bar{e}\in P_{rv}} a_v \right) \geq \sum_{v:f\in P_{rv}} a_v,$$

where $\bar{E}$ contains edges incident on j except e and f. Hence if $\alpha_f^r = \alpha_0$ then also $\alpha_e^r = \alpha_0$, otherwise

$$\alpha_f^r \le \min\{ \sum_{v:e \in P_{rv}} a_v, \alpha_0 \} = \alpha_e^r.$$

□

In the following theorem we claim that the strength of knapsack tree inequalities depends on the position of root r in the underlying tree. We show how to select r so that the best possible reduction of the coefficients and thus the strongest knapsack tree inequality for a given tree can be achieved.

Theorem 4.7 *Let* $T = (V_T, E_T)$ *be a tree in* G. *The strongest knapsack tree inequality defined on* T *is obtained for a root*

$$r \in \mathcal{R} := \mathrm{Argmin}_{v \in V_T} \sum_{e \in E_T} \alpha_e^v,$$

i.e., if $r \in \mathcal{R}$ *then*

$$\sum_{e \in E_T} \alpha_e^s y_e \ge \sum_{e \in E_T} \alpha_e^r y_e \ge \alpha_0 \tag{4.12}$$

holds for all $s \in V_T$ *and all* $y \ge 0$. *In particular,*

$$\sum_{e \in E_T} \alpha_e^r y_e = \sum_{e \in E_T} \alpha_e^s y_e \tag{4.13}$$

holds for all $r, s \in \mathcal{R}$ *and all* $y \ge 0$.

Proof. Let $\Pi = (V_\Pi, E_\Pi)$ be the path joining nodes $r \in \mathcal{R}$ and $s \in V_T$ with $V_\Pi = \{v_1, \ldots, v_p\}$, $p \ge 2$, where $v_1 = r$, $v_p = s$, and v_k, v_{k+1}, $1 \le k \le p-1$, are adjacent in T. Applying recursively Lemma 4.5 (1) to nodes v_i, v_{i+1}, $i = 1, \ldots, p-1$, we obtain

$$\alpha_e^r = \alpha_e^s, \qquad \forall e \in E_T \setminus E_\Pi. \tag{4.14}$$

By Lemma 4.6 we have

$$\begin{aligned}
\alpha_{rv_2}^r &\ge \alpha_{v_2 v_3}^r \ge \ldots \ge \alpha_{v_{p-2} v_{p-1}}^r \ge \alpha_{v_{p-1} s}^r, \\
\alpha_{rv_2}^s &\le \alpha_{v_2 v_3}^s \le \ldots \le \alpha_{v_{p-2} v_{p-1}}^s \le \alpha_{v_{p-1} s}^s.
\end{aligned} \tag{4.15}$$

Since r is a minimal root,

$$\sum_{e \in E_T} \alpha_e^r \le \sum_{e \in E_T} \alpha_e^{v_2}$$

holds. Applying Lemma 4.5 (2) to nodes r and v_2 and Lemma 4.5 (1) to nodes v_2 and s we obtain

$$\alpha_{rv_2}^r \le \alpha_{rv_2}^{v_2} = \alpha_{rv_2}^s.$$

This together with (4.15) yields

$$\alpha_e^r \leq \alpha_e^s, \quad \forall e \in E_\Pi. \tag{4.16}$$

Hence for each $e \in E_\Pi$ the inequality $(\alpha_e^r - \alpha_e^s) y_e \leq 0$ is trivially valid for $P_\mathcal{B}$ and from (4.14) we obtain

$$\sum_{e \in E_T} \alpha_e^r y_e - \sum_{e \in E_T} \alpha_e^s y_e = \sum_{e \in E_\Pi} (\alpha_e^r - \alpha_e^s) y_e \leq 0.$$

Thus (4.12) holds. The equation (4.13) follows directly from (4.12). □

The relation (4.16) in the above proof implies the following statement.

Remark 4.8 *Let $T = (V_T, E_T)$ be a tree in G and $r, s \in V_T$. Then the inequality*

$$\sum_{e \in E_T} \min\{\alpha_e^r, \alpha_e^s\} y_e \geq \alpha_0$$

is valid for $P_\mathcal{B}$.

In the sequel we call the elements of the set $\mathcal{R}$ **minimal roots** of a given tree T. In Theorem 4.7 we showed that all minimal roots of T lead to the same knapsack tree inequality and thus to obtain the strongest one it is sufficient to identify any minimal root. Assume we are given a tree $T = (V_T, E_T)$ rooted at some node r. In order to find a minimal root one can proceed iteratively as follows. Select a node $s \in V_T$ adjacent to r such that

$$\alpha_{rs}^r = \max\{\alpha_{rv}^r : rv \in E_T\}.$$

If $\alpha_{rs}^r > \alpha_{rs}^s$ then also

$$\sum_{e \in E_T} \alpha_e^r > \sum_{e \in E_T} \alpha_e^s,$$

by Lemma 4.5 (2). Hence r can be discarded and s is marked as root of T. Otherwise

$$\sum_{e \in E_T} \alpha_e^r y_e \geq \alpha_0$$

is the strongest knapsack tree inequality with respect to all possible choices of roots in T. The following propositions show that our strengthening procedure ends up with correct results. Due to Proposition 4.9 it is sufficient to search in the direct neighborhood of the current root for a possible improvement. Proposition 4.10 ensures that it is enough to examine some node adjacent to r maximizing α_{rv}^r, $rv \in E_T$.

Proposition 4.9 *Given a tree $T = (V_T, E_T)$ in G. Node r is a minimal root of T if and only if $\alpha_{rv}^r \leq \alpha_{rv}^v$ holds for all $v \in V_T$ adjacent to r.*

Proof. Let v be a node adjacent to r. We assume first that $r \in \mathcal{R}$. Then

$$\sum_{e \in E_T} \alpha_e^r \leq \sum_{e \in E_T} \alpha_e^v$$

holds and $\alpha_{rv}^r \leq \alpha_{rv}^v$ due to Lemma 4.5 (2). Now, assume that

$$\alpha_{rv}^r \leq \alpha_{rv}^v, \quad \forall rv \in E_T. \tag{4.17}$$

We are going to show that this implies

$$\sum_{e \in E_T} \alpha_e^r \leq \sum_{e \in E_T} \alpha_e^s$$

for any $s \in V_T$ and thus r is a minimal root. Using (4.17) and Lemma 4.5 (2) we obtain that

$$\sum_{e \in E_T} \alpha_e^r \leq \sum_{e \in E_T} \alpha_e^v$$

holds for all v adjacent to r. Now, let s be a node not adjacent to r. Similarly to the proof of Theorem 4.7 we consider a path Π joining r and s and derive relations (4.14) and (4.15). We apply next (4.17) to (4.15) to obtain (4.16). From (4.14) and (4.16) follows that

$$\sum_{e \in E_T} \alpha_e^r \leq \sum_{e \in E_T} \alpha_e^s.$$

$\square$

Proposition 4.10 *Given a tree* $T = (V_T, E_T)$ *in* G *and adjacent nodes* $r, s \in V_T$ *such that* $\alpha_{rs}^s < \alpha_{rs}^r$. *Then*

$$\sum_{e \in E_T} \alpha_e^s < \sum_{e \in E_T} \alpha_e^v$$

holds for all nodes $v \in V_T$ *such that* $v \neq s$ *and* v *is adjacent to* r. *Furthermore,*

$$\alpha_{rs}^r = \max\{\alpha_{rv}^r : rv \in E_T\}.$$

Proof. Let $v \neq s$ be a node adjacent to r. By Lemma 4.5 (1) $\alpha_e^s = \alpha_e^v$ holds for all $e \in E_T \setminus \{sr, rv\}$. Furthermore, we obtain the chain of inequalities

$$\alpha_{rv}^s \leq \alpha_{rs}^s < \alpha_{rs}^r = \alpha_{rs}^v \leq \alpha_{rv}^v,$$

where the first inequality follows from (4.11), the second from the assumption of this lemma, the equality from Lemma 4.5 (1) applied to nodes r, v and the last inequality again from (4.11). Thus

$$\sum_{e \in E_T} \alpha_e^s < \sum_{e \in E_T} \alpha_e^v.$$

From the relations

$$\alpha_{rv}^r = \alpha_{rv}^s \leq \alpha_{rs}^s < \alpha_{rs}^r, \quad v \neq s,$$

we obtain the second claim of the lemma, where the first equality follows from Lemma 4.5 (1) applied to r and s. $\square$

4.2 Sufficient Conditions to Induce Facets

Assume that $G = (V, E)$ is a tree with unique node weights, i.e., $f_i = 1$ for all $i \in V$. Consider the knapsack inequality

$$\sum_{i \in V} x_i \leq u_\tau \tag{4.18}$$

defining the knapsack polytope

$$\text{conv}\{x \in \{0,1\}^n : \sum_{i \in V} x_i \leq u_\tau\}.$$

Then the knapsack tree inequality (4.1) defined on the tree G with respect to the knapsack inequality (4.18) takes the form

$$\sum_{i \in V} (1 - \sum_{e \in E_{ri}} y_e) \leq u_\tau.$$

Applying the strengthening (4.2) and notation (4.3) yields

$$\alpha_0 = |V| - u_\tau = l_\tau,$$
$$\alpha_e^r = \min\{|V_e^r|, l_\tau\}, \quad \forall e \in E.$$

To distinguish our inequality from the general case (4.4) we set $\kappa_e^r := \alpha_e^r$ and consider

$$\sum_{e \in E} \kappa_e^r y_e \geq l_\tau \tag{4.19}$$

or $(\kappa^r)^T y \geq l_\tau$ for short.

Remark 4.11 *If G is a star centered at r then $(\kappa^r)^T \geq l_\tau$ coincides with the star inequality (2.6).*

In this section we are going to show that the assumption on r to be a minimal root is not only a necessary condition as it follows from Theorem 4.7 but also sufficient for $(\kappa^r)^T y \geq l_\tau$ to be facet-defining for the polytope P_B if the underlying graph is a tree and all nodes have equal weights. In fact in some cases, which we specify later, an additional condition is needed. For its formulation we introduce the term **branch-less path**, which is a path whose inner nodes are of degree 2.

Theorem 4.12 *Assume that $G = (V, E)$ is a tree rooted at a node $r \in V$, $|V| \geq 3$, $f_i = 1$ for all $i \in V$ and $1 \leq l_\tau < u_\tau$. The knapsack tree inequality $(\kappa^r)^T y \geq l_\tau$ is facet-defining for P_B if and only if one of the following conditions is satisfied*

(a) *$l_\tau > 1$, r is a minimal root and each branch-less path in G contains less than u_τ nodes or exactly u_τ nodes and one end-edge of this path is a leaf in G,*

(b) $l_\tau = 1$.

Remark 4.13 *Due to Corollary 2.13 the assumptions about the graph G in Theorem 4.12 guarantee that P_B is full-dimensional.*

Remark 4.14 *In the case of $l_\tau = 0$ ($\tau = 1$) the knapsack inequality $\sum_{v \in V} x_v \leq u_\tau$ is redundant for P_K and thus the corresponding knapsack tree inequalities for P_B.*

If T coincides with a branch-less path then either $|V| = u_\tau$ and thus $l_\tau = 0$, or, T contains a branch-less path having more than u_τ nodes. Therefore only condition (b) in Theorem 4.12 allows that T coincides with a branch-less path.

If $l_\tau = 1$ then each node in the tree G is a minimal root.

The proof of Theorem 4.12 is rather complex. It requires distinction of several cases depending on the degree of r as well as on coefficients of variables corresponding to edges on the branch-less paths to r. To keep the proof of Theorem 4.12 compact and easy to follow, we provide first several results which we will repeatedly use during the case distinction. After giving some definitions we specify properties of cuts, whose incidence vectors are roots of the knapsack tree inequality. Next, we list a few consequences of the assumption about the root of the tree. Finally we exploit the assumption about branch-less paths in the tree. In all lemmas presented below we assume that $G = (V, E)$ is a tree rooted at a node $r \in V$, $|V| \geq 2$, $f_i = 1$ for all $i \in V$ and $1 < l_\tau \leq u_\tau{}^1$. The item (b) in Theorem 4.12 will be handled separately.

4.2.1   Definitions

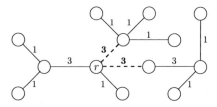

Figure 4.5:   Knapsack weights and the reduced knapsack weight (dashed) for $l_\tau = 3$.

To emphasize the assumption on the graph to be a tree we set $T := G$. For ease of exposition we call κ_e^r **knapsack weight** of $e \in E$ with respect to r. If $\kappa_e^r = l_\tau$ and $l_\tau < |V_e^r|$ we say that e has the **reduced knapsack weight**.

---

[1]In Section 4.2.7 we complement the assumption of Theorem 4.12 with the equipartition case.

Given a bipartition of V we denote by V_r the cluster containing root r and by $\bar{V}_r$ the complement of V_r in V. Each bipartition $(V_r, \bar{V}_r)$ corresponds to a unique cut $\Delta(V_r)$ and vice versa, since the underlying tree T is a connected graph.

A path in T consisting of one edge and its end-nodes is considered as a trivial branch-less path.

We say that two edges $e, f \in E$ are **related** if there exists a path to the root containing both e and f. An edge e is related to itself. For an edge e, the set

$$B_e = \{f \in E : f \text{ is related to } e\}$$

is called **branch** (induced by e). We say that a branch is **incident** on a node v if e is incident on v and $\deg(v)=1$ with respect to edges in B. If any two edges e and f are adjacent and related and such that f is closer to root than e (with respect to the number of edges) then f is **father** of e and e is **son** of f. We denote by S_e the set of all sons of an edge e. We refer to Appendix A.1 for other terms we use concerning nodes and edges of the tree T.

Below we put together a few relations following from the definition of knapsack weights and related edges.

Proposition 4.15 *Let r be the root of T and $e, f \in E$.*

(1) *If e and f are not related then $V_e^r \cap V_f^r = \emptyset$. Furthermore, if e and f are related and f lies on the path joining r and e then $V_e^r \subset V_f^r$.*

(2) *If f lies on the path joining r and e then $\kappa_e^r \leq \kappa_f^r$ holds. If $\kappa_e^r = l_r$ then $\kappa_e^r = \kappa_f^r$, otherwise $\kappa_e^r < \kappa_f^r$. In particular, if e is related to an edge with non-reduced knapsack weight, then the knapsack weight of e is also not reduced.*

(3) *Let S_e be the set of sons of e. If e does not have the reduced knapsack weight then*

$$\kappa_e^r = 1 + \sum_{\bar{e} \in S_e} \kappa_{\bar{e}}^r. \tag{4.20}$$

(4) *Let $B = (V_B, E_B)$ be a branch in T incident on r. Let*

$$\kappa_{\max} = \max_{e \in E_B} \kappa_e^r.$$

For each $k = 1, \ldots, \kappa_{\max}$ there exists a subset $D \subseteq E_B$ of non-related edges such that

$$\sum_{e \in D} \kappa_e^r = k.$$

Proof. (1) Follows directly from the definitions.

(2) Follows directly from Lemma 4.6 together with the assumption that $a_v = 1$, $v \in V$, and item (1).

(3) Let S_e be the set of sons of e and $i \in V$ the node incident on e and all $\bar{e} \in S_e$. Since all edges in S_e are not related, $V_{\bar{e}}^r \cap V_{\tilde{e}}^r = \emptyset$ holds for any two $\bar{e}, \tilde{e} \in S_e$, $\bar{e} \neq \tilde{e}$, due to (1). We have

$$V_e^r = \{i\} \,\dot{\cup}\, \bigcup_{\bar{e} \in S_e} V_{\bar{e}}^r.$$

By the assumption $\kappa_{\bar{e}}^r = |V_{\bar{e}}^r|$ and by (2) all edges in S_e also do not have reduced knapsack weights. Thus

$$\kappa_e^r = |V_e^r| = |\{i\}| + \sum_{\bar{e} \in S_e} |V_{\bar{e}}^r| = 1 + \sum_{\bar{e} \in S_e} \kappa_{\bar{e}}^r$$

holds.

(4) Let $f \in \mathrm{Argmax}\,\kappa_e^r$. Let v be the end-node of f such that $v \in V_f^r$. For $k = \kappa_{\max}, \dots, 1$ we reduce the number of nodes in V_f^r to obtain a set $\bar{V} \subset V_f^r$ with $|\bar{V}| = k$ and $|\bar{V}| = \dot{\bigcup}_{e \in D} V_e^r$. At the beginning we set $\bar{V} := V_f^r$. Obviously $|\bar{V}| \geq k$. First we check if $|\bar{V}| = k$. If not, we remove one node at a time adjacent to a son of f. If all nodes adjacent to sons of f are removed from $\bar{V}$, the nodes adjacent to the sons of a son of f are removed. We proceed iteratively in this way till $|\bar{V}| = k$.   □

4.2.2   Bisection Cuts Tight for Knapsack Tree Inequality

In the proof of Theorem 4.12 we consider bisection cuts whose incident vectors lie on the face of P_B induced by $(\kappa^r)^T y \geq l_\tau$. In this section we derive some elementary properties of these cuts.

Let $(V_r, \bar{V}_r)$ be a bipartition of V. The cut $\Delta(V_r)$ is **tight** for $(\kappa^r)^T y \geq l_\tau$ if its incidence vector $\chi^{\Delta(V_r)}$ satisfies

$$\sum_{e \in E} \kappa_e^r \chi_e^{\Delta(V_r)} = l_\tau.$$

Since $(V_r, \bar{V}_r)$ and its associated cut $\Delta(V_r)$ are in one-to-one correspondence, we also say that the bipartition $(V_r, \bar{V}_r)$ is tight for $(\kappa^r)^T y \geq l_\tau$.

A cut $\Delta(V_r)$, or a bipartition $(V_r, \bar{V}_r)$, is **double-tight** for $(\kappa^r)^T y \geq l_\tau$ if it is tight for $(\kappa^r)^T y \geq l_\tau$ and $\Delta(V_r)$ contains only edges with non-reduced knapsack weights. As we will show soon, if a bisection cut $\Delta(V_r)$ tight for $(\kappa^r)^T y \geq l_\tau$ contains only edges with non-reduced knapsack weights then $|V_r| = u_\tau$ and $|\bar{V}_r| = l_\tau$ hold. Hence the bisection $(V_r, \bar{V}_r)$ is tight with respect to the bounds u_τ and l_τ. That is why we call such $\Delta(V_r)$ or $(V_r, \bar{V}_r)$ double-tight for $(\kappa^r)^T y \geq l_\tau$. Below we list some obvious properties of bisection cuts tight for $(\kappa^r)^T y \geq l_\tau$.

Proposition 4.16 *Let $(V_r, \bar{V}_r)$ be tight for $(\kappa^r)^T y \geq l_\tau$.*

(1) $\Delta(V_r) = \{e\}$ *for some* $e \in E$ *if and only if* $\kappa_e^r = l_\tau$.

(2) *If* $|\Delta(V_r)| > 1$ *then it is double-tight for* $(\kappa^r)^T y \geq l_\tau$.

(3) *If* $\Delta(V_r) = \{e\}$ *for some* $e \in E$ *with the reduced knapsack weight then*

$$|\bar{V}_r| = |V_e^r| > l_\tau, \quad i.e., \quad |V_r| < u_\tau.$$

Lemma 4.17 *Any two edges in a bisection cut tight for* $(\kappa^r)^T y \geq l_\tau$ *are not related.*

Proof. Let T be rooted at $r \in V$ and $\Delta(V_r)$ be a bisection cut tight for $(\kappa^r)^T y \geq l_\tau$. W.l.o.g., we assume that $|\Delta(V_r)| > 1$. Let Δ_{even} be the set of edges with an even distance to r (with respect to the number of cut edges on the path to r) and $\Delta_{odd} = \Delta(V_r) \backslash \Delta_{even}$. Since $\Delta(V_r)$ is tight for $(\kappa^r)^T y \geq l_\tau$, we have, on the one hand,

$$\sum_{e \in \Delta_{odd}} \kappa_e^r + \sum_{e \in \Delta_{even}} \kappa_e^r = l_\tau.$$

On the other side, using the facts that if an edge belongs to a cut then its end-nodes belong to different clusters and that the total weight of $\bar{V}_r$ cannot fall below l_τ we obtain

$$\sum_{e \in \Delta_{odd}} \kappa_e^r - \sum_{e \in \Delta_{even}} \kappa_e^r \geq l_\tau.$$

Both relations can be satisfied only if $\sum_{e \in \Delta_{even}} \kappa_e^r$ vanishes. Hence $\Delta_{even} = \emptyset$ and its construction indicates that each path to r can be cut only once. $\square$

Corollary 4.18 *Let* $r \in V$ *be the root of* T. *For any bisection tight for inequality* $(\kappa^r)^T y \geq l_\tau$ *the subgraph of* T *induced by* V_r *is connected.*

Lemma 4.19 *Let* T *be rooted at* $r \in V$. *A bisection* $(V_r, \bar{V}_r)$ *is double-tight for* $(\kappa^r)^T y \geq l_\tau$ *if and only if* $|V_r| = u_\tau$ *and the subgraph* $(V_r, E(V_r))$ *is connected.*

Proof. We assume first that $(V_r, \bar{V}_r)$ is double-tight for $(\kappa^r)^T y \geq l_\tau$. By Corollary 4.18 any two edges in $\Delta(V_r)$ are not related, i.e., the subgraph $(V_r, E(V_r))$ is connected and

$$\bar{V}_r = \bigcup_{e \in \Delta(V_r)} V_e^r \quad \text{with} \quad V_e^r \cap V_f^r = \emptyset, \ \forall e, f \in \Delta(V_r).$$

Furthermore, $\kappa_e^r = |V_e^r|$ holds for each $e \in \Delta(V_r)$ and we obtain

$$|\bar{V}_r| = |\bigcup_{e \in \Delta(V_r)} V_e^r| = \sum_{e \in \Delta(V_r)} \kappa_e^r = l_\tau.$$

This yields $|V_r| = u_\tau$.

Now, consider a bisection $(V_r, \bar{V}_r)$ such that the subgraph $(V_r, E(V_r))$ is connected and $|V_r| = u_\tau$ (i.e., $|\bar{V}_r| = l_\tau$). We show first that $\Delta(V_r)$ contains only edges with non-reduced knapsack weights. Assume on the contrary that $\Delta(V_r)$ contains an edge f with the reduced knapsack weight. Since $\kappa_f^r = l_\tau$, this is the only edge in $\Delta(V_r)$, otherwise $\Delta(V_r)$ could not be tight for $(\kappa^r)^T y \geq l_\tau$. Hence $\Delta(V_r) = \{f\}$ and $|\bar{V}_r| = |V_f^r| > l_\tau$ holds contradicting the assumption that $|V_r| = u_\tau$. To show that $\Delta(V_r)$ is tight for $(\kappa^r)^T y \geq l_\tau$, we use the assumption that $(V_r, E(V_r))$ is connected. We have

$$\sum_{e \in \Delta(V_r)} \kappa_e^r = \sum_{e \in \Delta(V_r)} |V_e^r| = \left| \bigcup_{e \in \Delta(V_r)} V_e^r \right| = |\bar{V}| = l_\tau.$$

Hence $\Delta(V_r)$ is double-tight for $(\kappa^r)^T y \geq l_\tau$.   □

Corollary 4.20 *Let $r \in V$ a minimal root of T. $|V_r| \geq |\bar{V}_r|$ holds for any bisection $(V_r, \bar{V}_r)$ tight for inequality $(\kappa^r)^T y \geq l_\tau$.*

Remark 4.21 *All results presented in this section can be easily extended to the general case, when the knapsack inequality (4.4) is defined on a subtree of some graph G and the node weights are not one.*

4.2.3 Minimal Roots

We provide now some results following from the assumption that T is rooted at a node in $\mathcal{R}$. In particular we show that each edge of the tree is included in some bisection cut tight for $(\kappa^r)^T \geq l_\tau$. Furthermore, we specify a procedure to obtain a new bisection cut lying on the face induced by $(\kappa^r)^T \geq l_\tau$ from a given one.

Proposition 4.22 *Assume that $|V| \geq 3$ and $1 < l_\tau \leq u_\tau$. If $r \in V$ is a minimal root of T then $\deg(r) > 1$.*

Proof. Let $\deg(r) = 1$ and let s be the only node adjacent to r. We have

$$|V_{rs}^r| = u_\tau + l_\tau - 1 \geq 2l_\tau - 1 > l_\tau,$$
$$|V_{rs}^s| = 1 < l_\tau.$$

Hence $\kappa_{rs}^r = l_\tau > 1 = \kappa_{rs}^s$ and r cannot be a minimal root due to Proposition 4.9.   □

In the next lemma we give an upper bound on the total node weight of each branch incident on a minimal root.

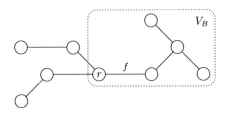

Figure 4.6: Node set V_B of a branch incident on r.

Lemma 4.23 *Let $B = (V_B, E_B)$ be a branch in T incident on a node $r \in V$. If r is a minimal root of T then $|V_B \setminus \{r\}| \leq u_r$.*

Proof. Let $B = (V_B, E_B)$ be a branch in T incident on r, see Figure 4.6, and assume that $|V_B \setminus \{r\}| > u_r$. Let s be the node in V_B adjacent to r and let $f = rs \in E_B$, see Figure 4.7. On the one hand, we have

$$\kappa_f^r = \min\{|V_f^r|, l_r\} = \min\{|V_B \setminus \{r\}|, l_r\} = l_r,$$

since $V_f^r = V_B \setminus \{r\}$ and $|V_B \setminus \{r\}| > u_r \geq l_r$. On the other hand,

$$\kappa_f^s = \min\{|V_f^s|, l_r\} = \min\{|V \setminus V_f^r|, l_r\} < l_r,$$

due to the fact that

$$|V \setminus V_f^r| = |V| - |V_f^r| < u_r + l_r - u_r = l_r.$$

Hence $\kappa_f^s < \kappa_f^r$ and r cannot be a minimal root by Proposition 4.9. $\qquad\square$

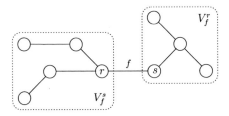

Figure 4.7: Node sets V_f^r and V_f^s.

The following lemma ensures the existence of all bisection cuts which we consider in the sufficiency proof of Theorem 4.12 in Section 4.2.5. In other words, if T is rooted at some node in $\mathcal{R}$ then for any edge in E there exists a bisection cut tight for $(\kappa^r)^T y \geq l_r$ containing this edge.

Lemma 4.24 *Let $r \in V$. For each $e \in E$ there exists a bisection cut tight for $(\kappa^r)^T y \geq l_r$ and containing e if and only if r is a minimal root in T.*

Proof. Note that since P_B is full-dimensional, see Corollary 2.13, for each $e \in E$ there exists a bisection cut containing e. First we assume that r is not a minimal root and find an edge that is not included in any bisection cut tight for $(\kappa^r)^T y \geq l_r$. By Proposition 4.9 there exists a node s adjacent to r such that $\kappa^r_{rs} > \kappa^s_{rs}$. Therefore $\kappa^s_{rs} = |V^s_{rs}| < l_r$. Let $\Delta(V_r)$ be a cut containing edge rs and assume first that $\kappa^r_{rs} = l_r$. If $|\Delta(V_r)| > 1$ then

$$\sum_{f \in \Delta(V_r)} \kappa^r_f = \kappa^r_{rs} + \underbrace{\sum_{f \in \Delta(V_r) \setminus \{rs\}} \kappa^r_f}_{> 0} > l_r.$$

Therefore $\Delta(V_r)$ is not tight for $(\kappa^r)^T y \geq l_r$. In case $\Delta(V_r) = \{rs\}$ we have $V_r = V^s_{rs}$ and hence $(V_r, \bar{V}_r)$ is not a bisection. If $\kappa^r_{rs} < l_r$ then $|V^r_{rs}| < l_r$ and we obtain the following contradiction

$$l_r + u_r = |V| = |V^s_{rs}| + |V^r_{rs}| < 2l_r \leq l_r + u_r.$$

Now, assume that r is a minimal root of T. We distinguish two cases:

(1) κ^r_e is reduced.

Then $\kappa^r_e = l_r$ and the cut $\Delta(V_r) = \{e\}$ is obviously tight for $(\kappa^r)^T y \geq l_r$. Assume that $\Delta(V_r)$ is not a bisection cut. Then either $|V_r| < l_r$ or $|\bar{V}_r| < l_r$. In the first case, let s be a node incident on e such that the path $\Pi_{rs} = (V_{rs}, E_{rs})$ joining r and s contains e. For $f \in E_{rs}$ we have by Lemma 4.6

$$\kappa^r_f \geq \kappa^r_e = l_r > |V_r| \geq |V^s_f| = \kappa^s_f,$$

see Figure 4.8.

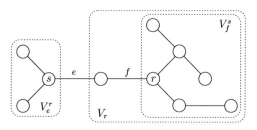

Figure 4.8: Node sets V_r, V^r_e and V^r_f considered in case (1).

From Lemma 4.5 (1) it follows that $\kappa^r_f = \kappa^s_f$ for all $f \in E_T \setminus E_{rs}$. Hence $\sum_e \in E_T \kappa^s_e < \sum_{e \in E_T} \kappa^r_e$ contradicts the assumption that r is a minimal root. The case $|\bar{V}_r| < l_r$ is not possible as well, because then we would have

$$\kappa^r_e = \min\{|V^r_e|, l_r\} = |V^r_e| = |\bar{V}_r| < l_r$$

against the assumption $\kappa^r_e = l_r$.

(2) κ_e^r is not reduced, i.e., $|V_e^r| \le l_\tau$.

This yields $|V \setminus V_e^r| \ge u_\tau$ and hence there exists a set $V_1 \subseteq V \setminus V_e^r$ such that the subgraph $(V_1, E(V_1))$ is connected and $|V_1| = u_\tau$. Thus the bipartition $(V_1, V \setminus V_1)$ is a bisection. For convenience we set $\bar{V}_1 := V \setminus V_1$. Now, we can select V_1 so that $e \in \Delta(V_1)$. In case $r \in V_1$ we obtain by Lemma 4.19 that $\Delta(V_1)$ is double-tight for $(\kappa^r)^T y \ge l_\tau$. Hence we get a bisection cut containing e and tight for $(\kappa^r)^T y \ge l_\tau$.

It turns out that $r \in V_1$ is granted by the assumption that r is a minimal root, as we show below. Assume that $r \notin V_1$. Note that $r \notin V_e^r$ as well. Hence $r \in \bar{V}_1 \setminus V_e^r$.

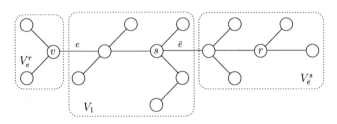

Figure 4.9: Node sets V_1, V_e^r and $V_{\bar{e}}^s$.

Let v be a node incident on e such that the path $\Pi_{rv} = (V_{rv}, E_{rv})$ joining r and v contains e, see Figure 4.9. We have:

(a) r and v are in one cluster, $(r, v \in \bar{V}_1)$;

(b) $e \in E_{rv} \cap \Delta(V_1)$;

(c) the subgraph $(V_1, E(V_1))$ is connected.

This yields $|E_{rv} \cap \Delta(V_1)| = 2$. We select $\bar{e} \in E_{rv} \cap \Delta(V_r)$ such that $\bar{e} \ne e$. Let $s \in V_1$ be a node incident on $\bar{e}$ such that the path $\Pi_{rs} = (V_{rs}, E_{rs})$ joining r and s contains $\bar{e}$, see Figure 4.9. On the one hand, since $V_{\bar{e}}^s = \bar{V}_1 \setminus V_e^r$

$$l_\tau > |V_f^s| = \kappa_f^s, \qquad \forall f \in E_{rs}, \tag{4.21}$$

holds. On the other hand, $V_1 \cup V_e^r \subseteq V_f^r$ and hence $|V_f^r| > u_\tau \ge l_\tau$ for all $f \in E_{rs}$ and we obtain

$$\kappa_f^r = \min\{|V_f^r|, l_\tau\} = l_\tau, \qquad \forall f \in E_{rs}. \tag{4.22}$$

By (4.21) and (4.22) we have

$$\kappa_f^r > \kappa_f^s, \qquad \forall f \in E_{rs}.$$

Thus r cannot be a minimal root due to Proposition 4.9. $\square$

The following observations we obtain from the previous proof.

Corollary 4.25 *Let $r \in V$ and $\kappa_e^r = l_\tau$ for some $e \in E$. If r is a minimal root of T then $\Delta(V_e^r) = \{e\}$ is a bisection cut.*

Corollary 4.26 *Let $r \in V$ and let e be an edge in E with a non-reduced knapsack weight. If r is a minimal root then each bisection cut containing e which is tight for $(\kappa^r)^T y \geq l_r$ is double-tight.*

Corollary 4.27 *Let $B = (V_B, E_B)$ be a branch in T incident on a minimal root r and let $\bar{E} \subseteq E_B$ contain non-related edges such that*

$$\sum_{e \in \bar{E}} \kappa_e^r \leq l_r.$$

Then there exists a bisection cut which contains $\bar{E}$ and is double-tight for $(\kappa^r)^T y \geq l_r$.

Proof. The above statement is an extension to case (2) in the proof of Lemma 4.24. Here we consider $\bigcup_{e \in \bar{E}} V_e^r$ instead of V_e^r and a set $V_1 \subset V \setminus (\bigcup_{e \in \bar{E}} V_e^r)$. By a similar argument as in the proof of Lemma 4.24 one obtains the existence of the desired bisection cut.   □

Corollary 4.28 *Let $B = (V_B, E_B)$ be a branch in T incident on r with $\deg(r) = 3$ and let $\bar{E} \subseteq E_B$ contain non-related edges such that*

$$\sum_{e \in \bar{E}} \kappa_e^r \leq l_r - 2. \tag{4.23}$$

There exists a double-tight bisection $(V_r, \bar{V}_r)$ for $(\kappa^r)^T y \geq l_r$ such that $\Delta(V_r) \cap E_B = \bar{E}$ and $\bar{V}_r \setminus V_B$ contains at least two non-adjacent nodes.

Proof. From the assumption (4.23) and $\Delta(V_r) \cap E_B = \bar{E}$ follows that $|\bar{V}_r \setminus V_B| \geq 2$. Let $s, t \in \bar{V}_r \setminus V_B$. Assume that for any selection of the set V_r, which fulfills the assumptions of the corollary, s and t are the only nodes in $\bar{V}_r \setminus V_B$ and are adjacent. This is only possible if the subgraph induced by $(V \setminus V_B) \cup \{r\}$ is a path in T, i.e., $\deg(r) = 2$.   □

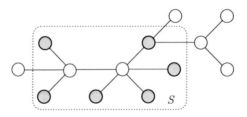

Figure 4.10: Boundary ∂S (gray nodes) of node subset S.

For convenience we introduce the following term. Let $S \subseteq V$ such that the subgraph $(S, E(S))$ of T induced by S is connected. A node $v \in S$ is called a **boundary node** if the subgraph induced by $S \setminus \{v\}$ is connected. The subset of all boundary nodes of S is called the **boundary** of the node set S and denoted by ∂S, see Figure 4.10.

The next lemma provides a tool which we will often use to obtain a new bisection or bisection cut tight for $(\kappa^r)^T y \geq l_\tau$ from a given one.

Lemma 4.29 *Assume that $|V| \geq 3$ and $1 < l_\tau \leq u_\tau$. Let r be a root of T. Let $(V_r, \bar{V}_r)$ be a double-tight bisection for $(\kappa^r)^T y \geq l_\tau$.*

(1) *For each $s \in \bar{V}_r$ adjacent to a node $u \in \partial V_r$ there exists a bisection $(V', \bar{V}')$ such that*
$$V' = \{s\} \cup V_r \setminus \{v\} \quad and \quad \bar{V}' = \{v\} \cup \bar{V}_r \setminus \{s\}$$
for some $v \in \partial V_r$, $v \neq u$, see Figure 4.11.

(2) *If $v \neq r$ then the bisection cut $\Delta(V')$ is double-tight for $(\kappa^r)^T y \geq l_\tau$.*

(3) *If $v = r$ and r is minimal then $\Delta(V')$ is tight for $(\kappa^r)^T y \geq l_\tau$.*

(4) *If $v = r$, r is minimal, and $\tau = 0$ then $\Delta(V')$ is double-tight for $(\kappa^r)^T y \geq l_\tau$.*

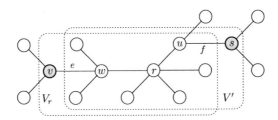

Figure 4.11: Swapping v for s.

Proof.

(1) Due to the definition of V' and $\bar{V}'$ we have $|V'| = |V_r|$ and $|\bar{V}'| = |\bar{V}_r|$. Hence $l_\tau \leq |V'|, |\bar{V}'| \leq u_\tau$ holds and $(V', \bar{V}')$ is also a bisection.

(2) If $v \neq r$ then $r \in V'$. Due to the definition the subgraph of T induced by V' is connected and $|V'| = u_\tau$. Hence $\Delta(V')$ is double-tight by Lemma 4.19.

(3) If $v = r$ then $r \in \bar{V}'$. Let $e = rw$. Since $v \in \partial V_r$, we obtain that $(V', E(V'))$ is a connected subgraph of T and in this particular case $V' \subseteq V_e^r$. Hence
$$|V_e^r| \geq |V'| = u_\tau \geq l_\tau. \tag{4.24}$$

Therefore $\kappa_e^v = l_\tau$ and $\Delta(V')$ is tight for $(\dot\kappa^r)^T y \geq l_\tau$ if it contains exclusively the edge e. Obviously $\Delta(V') = \{e\}$ is satisfied if also $(\bar V', E(\bar V'))$ is connected. Indeed, the latter assertion follows from the assumption that r is a minimal root of T. Namely, assume the contrary, then $V_e^w \subset \bar V'$ and

$$l_\tau = |\bar V'| > |V_e^w| = \kappa_e^w.$$

From (4.24) we get $\kappa_f^w < \kappa_f^r$ and since $\kappa_e^r = \kappa_e^w$ for $f \neq e$, due to Lemma 4.5 (1), r cannot be a minimal root.

(4) If $\tau = 0$ then $u_\tau = l_\tau$. By (3) $(V', E(V'))$ and $(\bar V', E(\bar V'))$ are connected. Hence $\kappa_e^r = l_\tau = u_\tau = |V_e^r|$ and therefore κ_e^r is not reduced. Furthermore, in this particular case $\Delta(V') = \{e\}$ is double-tight for $(\kappa^r)^T y \geq l_\tau$.   □

From the proof of Lemma 4.29 we obtain the following implications.

Corollary 4.30 *Let $(V', \bar V')$ be a bisection of T as defined in Lemma 4.29 (1).*

(1) $\Delta(V')$ *is a bisection cut tight for* $(\kappa^r)^T y \geq l_\tau$.

(2) *If $v = r$ and r is a minimal root of T then both $(V', E_{V'})$ and $(\bar V', E_{\bar V'})$ are connected.*

In Lemma 4.29 we obtained a new bisection $(V', \bar V')$ from a given bisection $(V_r, \bar V_r)$ by, as we will call it in the sequel, **swapping** a node $v \in \partial V_r$ for a node in $\bar V_r$. The subgraph induced by the set V' is connected and $|V'| = u_\tau$. Additionally, if $v \neq r$ the set V' inherits root r and in this case V' will be called **V_r-shift**.

Applying the swapping operation we obtain the following important relationship.

Corollary 4.31 *Let F and F_b be faces of P_B defined by $(\kappa^r)^T y \geq l_\tau$ and $b^T y \geq b_0$, respectively, such that $F \subseteq F_b$. Let $(V_r, \bar V_r)$ and $(V', \bar V')$ be bisections of T as defined in Lemma 4.29 such that $v \neq r$. Then*

$$b_f - \sum_{\bar e \in S_f} b_{\bar e} = b_e - \sum_{\bar e \in S_e} b_{\bar e} \tag{4.25}$$

holds, where S_e and S_f are sets of sons of edges $e = vw$ and $f = us$, respectively.

Proof. There holds $r \in V'$, since we assume that $v \neq r$. We have $\Delta(V_r) = \{f\} \cup S_e \cup D$ and $\Delta(V') = \{e\} \cup S_f \cup D$, see Figure 4.11. $\Delta(V_r)$ and $\Delta(V')$ are tight for $(\kappa^r)^T y \geq l_\tau$ by Lemma 4.29. Hence their incidence vectors lie on F_b. Therefore

$$b_0 = \sum_{e \in \Delta(V_r)} b_e = \sum_{e \in \Delta(V')} b_e$$

and

$$b_f + \sum_{\bar e \in S_e} b_{\bar e} + \sum_{\bar e \in D} b_{\bar e} = b_e + \sum_{\bar e \in S_f} b_{\bar e} + \sum_{\bar e \in D} b_{\bar e}.$$

The latter equation directly yields the relation (4.25).   □

4.2.4 Branch-less Paths

The next lemmas concern branch-less paths in T having at least u_τ nodes. We explain, why branch-less paths in T cannot be too long with respect to the number of nodes.

Since each sub-path of a branch-less path is a branch-less path, we introduce the following definition to distinguish the biggest path with respect to node inclusion.

A branch-less path $\Pi = (V_\Pi, E_\Pi)$ in a subgraph S of T is called **maximal** in S if

$$|V_\Pi| = \max\{|V_b| : (V_b, E_b) \text{ is a branch-less path in } S\},$$

i.e., it is the branch-less path in S of the biggest node cardinality.

First we present some connections between branch-less paths and minimal roots.

Proposition 4.32 *Let* $\Pi = (V_\Pi, E_\Pi)$ *be a branch-less path in* T.

(1) *If* $|V_\Pi| > u_\tau$ *then* V_Π *contains all minimal roots of* T.

(2) *If* Π *is maximal with* $|V_\Pi| = u_\tau$ *and none of end-nodes of* Π *is a leaf in* T *then all minimal roots of* T *are inner nodes of* Π.

(3) *If* $l_\tau > 0$, Π *is maximal with* $|V_\Pi| = u_\tau$ *and one of the end-edges of* Π *is a leaf in* T *then all minimal roots of* T *are nodes in* V_Π.

Proof.

(1) Let r be a minimal root of T. Assume $r \notin V_\Pi$ is adjacent to an end-node of Π, say s. On the one hand, we have $V_\Pi \subseteq V_{rs}^r$. Hence $|V_{rs}^r| > u_\tau \geq l_\tau$ yields $\kappa_{rs}^r = l_\tau$. On the other hand,

$$|V_{rs}^s| \leq |V| - |V_\Pi| < l_\tau.$$

Therefore $\kappa_{rs}^s < \kappa_{rs}^r$ and by Lemma 4.5 (2) r cannot be a minimal root of T. Using Lemma 4.6 one obtains similarly a contradiction if $r \notin V_\Pi$ and r is not adjacent to a node in V_Π.

(2) Let r be a minimal root of T and assume that r is an end-node of Π. Let s be the node in V_Π adjacent to r. Since the other end-node of Π is not a leaf, we have, on the one hand,

$$|V_{rs}^r| \geq u_\tau + 1 > l_\tau$$

yielding $\kappa_{rs}^r = l_\tau$. On the other hand, we have

$$|V_{rs}^s| = |V| - |V_{rs}^r| \leq |V| - u_\tau - 1 < l_\tau.$$

Therefore $\kappa_{rs}^s < \kappa_{rs}^r$ and again by Lemma 4.5 (1) r cannot be a minimal root of T.

(3) The assumption $l_\tau > 0$ forces that at least one end-edge of Π is not a leaf in T. Let node s be the end-node of this edge such that $\deg(s) = 1$ with respect to Π. Since Π is a maximal branch-less path, we have $\deg(s) > 2$ with respect to T. Assume that T is rooted at some $r \in V \setminus V_\Pi$ which is adjacent to s. On the one hand, we have

$$|V_{rs}^r| > |V_\Pi| = u_\tau \ge l_\tau.$$

Hence $\kappa_{rs}^r = l_\tau$. On the other hand,

$$|V_{rs}^r| < |V \setminus V_\Pi| = l_\tau$$

yields $\kappa_{rs}^s < l_\tau$. Again, by Lemma 4.5 (1) r cannot be a minimal root of T. Using Lemma 4.6 one obtains similarly a contradiction if $r \notin V_\Pi$ and r is not adjacent to a node in V_Π.   □

Proposition 4.33 *Assume that T is rooted at a minimal root r and $l_\tau > 0$. Furthermore, assume that T contains a maximal branch-less path $\Pi = (V_\Pi, E_\Pi)$ such that $|V_\Pi| = u_\tau$ and one of the end-edges of Π is a leaf in T. If all maximal branch-less paths to r contain edges with non-reduced knapsack weights then $\deg(r) > 2$.*

Proof. Due to the assumption $l_\tau > 0$ at least one end-edge of Π is not a leaf in T. By Proposition 4.32 (3) root r of T lies on the path Π. The only node in V_Π with degree greater than 2 is the end-node of Π, say s, which is not a leaf-node in T. Assume that $r \ne s$. Let e be the end-edge of Π incident on s. We have $V_e^r = (V \setminus V_\Pi) \cup \{s\}$, i.e., $|V_e^r| = l_\tau + 1$, and thus e has the reduced knapsack weight. The same holds for all edges preceding e on the path to r due to Proposition 4.15 (2). Therefore the maximal branch-less path to r containing edge e has only edges with reduced knapsack weight.
□

Lemma 4.34 *Let $\Pi = (V_\Pi, E_\Pi)$ be a branch-less path in a tree T rooted at a node r and let $(V_r, \bar{V}_r)$ be a bisection tight for $(\kappa^r)^T y \ge l_\tau$. If $|V_\Pi| \le u_\tau + 1$ then $|\Delta(V_r) \cap E_\Pi| \le 1$. If $|\Delta(V_r) \cap E_\Pi| = 2$ then $|V_\Pi| \ge u_\tau + 2$.*

Proof. Let $\Pi = (V_\Pi, E_\Pi)$ be a branch-less path in T and let $(V_r, \bar{V}_r)$ be a bisection tight for $(\kappa^r)^T y \ge l_\tau$. Due to Lemma 4.17 $|\Delta(V_r) \cap E_\Pi| \ge 2$ is only possible if root r of T lies on the path Π. Furthermore, if $|\Delta(V_r) \cap E_\Pi| = 2$ then these two edges are the only edges in $\Delta(V_r)$. Let $\Delta(V_r) = \{e, f\}$ with $e, f \in E_\Pi$. We have

$$\kappa_e^r + \kappa_f^r = l_\tau \quad \text{and} \quad 0 < \kappa_e^r, \, 0 < \kappa_f^r.$$

Hence $\kappa_e^r < l_\tau$, $\kappa_f^r < l_\tau$ and $\Delta(V_r)$ is double-tight for $(\kappa^r)^T y \ge l_\tau$. Note that V_r is the set of nodes on the path joining edges e and f. We have $V_r \subset V_\Pi$ and $|V_r| = u_\tau$ by Lemma 4.19. Since V_r contains at least 2 nodes less than V_Π, we obtain

$$|V_\Pi| - 2 \ge |V_r| = u_\tau,$$

i.e., $|V_\Pi| \ge u_\tau + 2$.   □

Proposition 4.35 *If T includes a branch-less path $\Pi = (V_\Pi, E_\Pi)$ having at least $u_\tau + 1$ nodes then all incidence vectors of bisection cuts tight for $(\kappa^r)^T y \geq l_\tau$ satisfy also the equality*

$$\sum_{e \in E'} y_e = 1 \qquad (4.26)$$

for any sub-path $\Pi' = (V', E')$ of Π such that $|V'| = u_\tau + 1$.

Proof. Let $\Pi' = (V', E')$ be a sub-path of Π having exactly $u_\tau + 1$ nodes. By Lemma 4.34

$$\sum_{e \in E'} y_e \leq 1$$

holds for all incidence vectors of bisection cuts tight for $(\kappa^r)^T y \geq l_\tau$. Since the total node weight of path Π' exceeds u_τ, it must be cut at least once. This means, a bisection $(V_r, \bar{V}_r)$ such that

$$\sum_{e \in E'} \chi_e^{\Delta(V_r)} = 0$$

does not exist and the claim of the lemma follows. □

Proposition 4.36 *Let T be rooted at $r \in V$. Assume that the maximal branch-less path $\Pi = (V_\Pi, E_\Pi)$ in T has exactly u_τ nodes, r is some inner node of Π, and both end-edges f, g of Π are not leaves in T. Then all incidence vectors of bisection cuts tight for $(\kappa^r)^T y \geq l_\tau$ satisfy also the equality*

$$n_2 \sum_{e \in D_1} \kappa_e^r y_e + n_1 \sum_{e \in E_1} (l_\tau - \kappa_e^r) y_e - n_1 \sum_{e \in D_2} \kappa_e^r y_e - n_2 \sum_{e \in E_2} (l_\tau - \kappa_e^r) y_e = 0, \qquad (4.27)$$

where $n_1 + n_2 = l_\tau$, $\Pi_1 = (V_1, E_1)$, $\Pi_2 = (V_2, E_2)$ are sub-paths of Π such that $E_1 \dot{\cup} E_2 = E_\Pi$ and $V_1 \cap V_2 = \{r\}$, D_1 and D_2 are sets of all edges in $E \setminus E_\Pi$ related to edges in Π_1 and Π_2, respectively, see Figure 4.12.

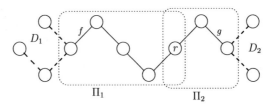

Figure 4.12: A maximal branch-less path in T having exactly u_τ (= 7) nodes.

Proof. Given the assumptions of the lemma regarding Π_1, Π_2, D_1, and D_2 as well as the notation in Figure 4.12. Let f and g be the end-edges of Π_1 and Π_2, respectively, not adjacent to r. Note that $D_1 = E(V_f^r)$ and $D_2 = E(V_g^r)$. We first show that $n_1 + n_2 = l_r$ holds for

$$n_1 = \sum_{e \in S_f} \kappa_e^r \quad \text{and} \quad n_2 = \sum_{e \in S_g} \kappa_e^r, \tag{4.28}$$

where S_f and S_g are the sets of sons of f and g, respectively. Since $r \in V_\Pi$, $|V_\Pi| = u_r$ and Π as a path is a connected graph, we obtain by Lemma 4.19 that $\Delta(V_\Pi) = S_f \cup S_g$ is (double-)tight for $(\kappa^r)^T y \geq l_r$, i.e.,

$$l_r = \sum_{e \in \Delta(V_\Pi)} \kappa_e^r = \sum_{e \in S_f} \kappa_e^r + \sum_{e \in S_g} \kappa_e^r = n_1 + n_2,$$

as we claimed. Note that $n_1 > 0$ and $n_2 > 0$ due to the assumption that neither f nor g is a leaf in T.

By Lemma 4.34 each bisection cut Δ tight for $(\kappa^r)^T y \geq l_r$ has at most one common edge with E_Π. We distinguish the following two cases:

(1) $\Delta \cap E_\Pi = \emptyset$.
    Here $V_r = V_\Pi$ and thus $\Delta(V_r) = S_f \cup S_g$. It can easily be checked that the incidence vector of $\Delta(V_r)$ satisfies (4.27) with n_1 and n_2 defined in (4.28).

(2) $|\Delta \cap E_\Pi| = 1$.
    Assume that $|\Delta \cap E_1| = 1$, i.e., $\Delta = \{e_1\} \cup \tilde{D}_2$ with $e_1 \in E_1$ and $\tilde{D}_2 \subseteq D_2$. Inserting χ^Δ into (4.27) we obtain

$$l_r - \kappa_{e_1}^r = \sum_{e \in \tilde{D}_2} \kappa_e^r,$$

which is satisfied due to the assumption that Δ is tight for $(\kappa^r)^T y \geq l_r$.
    Because of the symmetry of equation (4.27) the proof for the case $|\Delta \cap E_2| = 1$ is carried out analogously. $\qquad \square$

Corollary 4.37 *Given the assumptions of Propositions 4.35 and 4.36, respectively.*

(1) *(4.26) coincides with $(\kappa^r)^T y = l_r$ if and only if $l_r = 1$. In this case T is a path.*

(2) *(4.27) does note coincide with $(\kappa^r)^T y = l_r$.*

Proof.

(1) T must contain a branch-less path having at least $u_r + 1$ nodes to fulfill the assumptions of Proposition 4.35. Let $\Pi = (V_\Pi, E_\Pi)$ be a path in T such that $|V_\Pi| = u_r + 1$. The sequence of the following equivalent statements yields the claim. (4.26) coincides with $(\kappa^r)^T y = l_r$ if and only if

$$\sum_{e \in E} \kappa_e^r y_e = \sum_{e \in E_\Pi} y_e = 1$$

$$\Leftrightarrow \quad l_r = 1 \quad \text{and} \quad E = E_\Pi$$

$$\Leftrightarrow \quad |V| = u_r + 1$$

$$\Leftrightarrow \quad |V_\Pi| = |V|$$

$$\Leftrightarrow \quad \Pi = T.$$

(2) Since the right-hand side of (4.27) is equal to 0, this equality can coincide with $(\kappa^r)^T y = l_\tau$ only if $l_\tau = 0$. In this case $\kappa^r_e = 0$ for all $e \in E$ and the left hand side of (4.27) is trivially equal to 0. □

Corollary 4.38 *Assume that $P_\mathcal{B}$ is full-dimensional and $l_\tau > 1$. If there exists a branch-less path in T having either at least $u_\tau + 1$ nodes or exactly u_τ nodes and none of its end-edges is a leaf in T then $(\kappa^r)^T y \geq l_\tau$ does not define a facet of $P_\mathcal{B}$.*

Proof. Assume that T contains a branch-less path with number of nodes at least $u_\tau + 1$, or, exactly u_τ nodes and none of its end-edges is a leaf in T. By Propositions 4.35 and 4.36 each bisection cut tight for $(\kappa^r)^T y \geq l_\tau$ satisfies some equality which for $l_\tau > 1$ does not coincide with $(\kappa^r)^T y = l_\tau$, as we showed in Corollary 4.37. By Proposition A.2 the face defined by the inequality $(\kappa^r)^T y \geq l_\tau$ cannot be a facet of $P_\mathcal{B}$. □

4.2.5 Sufficiency Proof of Theorem 4.12

Let F_κ be the face of $P_\mathcal{B}$ induced by $(\kappa^r)^T y \geq l_\tau$ and F_b be a facet of $P_\mathcal{B}$ defined by an inequality $b^T y \geq b_0$ such that $F_\kappa \subseteq F_b$. We apply Proposition A.1 to prove that $(\kappa^r)^T y \geq l_\tau$ is a facet defining inequality for $P_\mathcal{B}$. In particular, we show that $F_\kappa = F_b$, i.e., there exists a $\gamma \in \mathbb{R} \setminus \{0\}$ such that

$$b_e = \gamma \kappa^r_e, \quad \forall e \in E, \tag{4.29}$$

$$b_0 = \gamma l_\tau.$$

To derive the above relations, i.e., the equivalence between the inequalities $b^T y \geq b_0$ and $(\kappa^r)^T y \geq l_\tau$, we consider incidence vectors of bisection cuts lying on the corresponding faces of $P_\mathcal{B}$. To obtain a sufficient number of bisection cuts we apply the swapping procedure. In particular, using relation (4.25) we show (4.29) step by step by proving the sequence of consecutive results. Generally we distinguish two cases: (A) there is at least one maximal branch-less path to the root in T whose edges all have only the reduced knapsack weight and (B) all maximal branch-less paths to the root contain edges with non-reduced knapsack weights. The latter case demands a subdivision into cases depending on the degree of the root of T.

The following lemma is a foundation for the proofs of propositions concerning cases (A) and (B).

Lemma 4.39 *Let F_κ and F_b be faces of $P_\mathcal{B}$ defined by $(\kappa^r)^T y \geq l_\tau$ and $b^T y \geq b_0$, respectively, such that $F_\kappa \subseteq F_b$. Let $B = (V_B, E_B)$ be a branch in T incident on $r \in V$ and $\Pi = (V_\Pi, E_\Pi)$ be the maximal branch-less path to r in B. Assume that r is a minimal root of T. Then there is a $\gamma_B \neq 0$ such that*

(1) $b_e = b_0$ holds for any $e \in E_T$ with $\kappa^r_e = l_\tau$,

(2) $b_e = b_f = \gamma_B$ holds for any two leaves $e, f \in B$,

(3) $b_e = \gamma_B \kappa_e^r$ holds for $e \in E_B \setminus E_\Pi$,

(4) $b_e = \gamma_B \kappa_e^r$ holds for $e \in E_B$ if E_Π contains only edges with the reduced knapsack weight.

Proof.

(1) Let e be some edge in E_T with $\kappa_e^r = l_\tau$. $\Delta(V_e^r) = \{e\}$ is a bisection cut tight for $(\kappa^r)^T y \geq l_\tau$ due to Corollary 4.25. Since $\chi^{\{e\}}$ lies on F_b, we obtain

$$b_e = b_0.$$

(2) Let e, f be leaves in E_B and s, v their respective end-nodes of degree 1. $|V_B \setminus \{s\}| \leq F$ follows from Lemma 4.23. Hence there exists a bisection $(V_r, \bar{V}_r)$ such that $V_B \setminus \{s\} \subseteq V_r$, $|V_r| = u_\tau$, and $(V_r, E(V_r))$ is connected, i.e., $(V_r, \bar{V}_r)$ is double-tight for $(\kappa^r)^T y \geq l_\tau$. Using Lemma 4.29 we swap s for v and get a new bisection $(V', \bar{V}')$, see Figure 4.13. Since $v \neq r$, we apply Corollary 4.31 and by (4.25) we obtain

$$b_e = b_f =: \gamma_B. \tag{4.30}$$

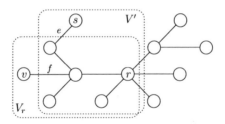

Figure 4.13: V_r-shift considered in the proof of case (2).

(3) If $B \subseteq \Pi$ there is nothing to prove, so assume that $\Pi \subset B$. Let $e \in E_B \setminus E_\Pi$. If e is a leaf then

$$b_e = \gamma_B = \gamma_B \kappa_e^r$$

follows from (2). Hence we assume now that e is not a leaf and consider further two cases:

(3a) κ_e^r is not reduced.
     Since $e \notin E_\Pi$, there exists a leaf $l \in E_B \setminus E_\Pi$ not related to e. Let S_e be the set of sons of e. By Lemma 4.6 all edges in S_e do not have the reduced knapsack weight. We have

$$\sum_{f \in S_e} \kappa_f^r + \kappa_l^r = \sum_{f \in S_e} |V_f^r| + |V_l^r| = |V_e^r| = \kappa_e^r \leq l_\tau$$

and by Corollary 4.27 there exists a bisection $(V_r, \bar{V}_r)$ such that the cut

$$\Delta(V_r) = S_e \cup \{l\} \cup D$$

is double-tight for $(\kappa^r)^T y \geq l_r$. D is a (possibly empty) set of edges related neither to e nor to l. Thus $|V_r| = u_r$ by Lemma 4.19. Let s be the end-node of l of degree 1 and v be the node adjacent to e and its sons, see Figure 4.14. We swap v for s and obtain a new bisection by Lemma 4.29. Since $v \neq r$, we

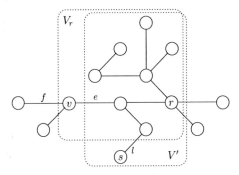

Figure 4.14: V_r-shift considered in case (3a) for $l_r = 6$.

use Corollary 4.31. By (4.25) and (4.30)

$$b_e = b_l + \sum_{f \in S_e} b_f = \gamma_B + \sum_{f \in S_e} b_f \qquad (4.31)$$

holds. If all edges in S_e are leaves then by (4.31) and case (2) we obtain

$$b_e = b_l + \sum_{f \in S_e} b_f = \gamma_B + \gamma_B |S_e| = \gamma_B \kappa_e^r. \qquad (4.32)$$

If e is from higher depth level of the tree, we apply recursively (4.31) and (4.32). In this way we obtain

$$b_e = \gamma_B \kappa_e^r$$

for all edges in $E_B \setminus E_{\mathrm{II}}$.

(3b) κ_e^r is reduced.

By Corollary 4.25 the cut $\Delta(V_e^r) = \{e\}$ is a bisection cut tight for $(\kappa^r)^T y \geq l_r$. Since κ_e^r is reduced, there exists a set, say S_e, of edges related to e (but not to each other) with the total knapsack weight equal to l_r, see Figure 4.15. Note that in this particular case S_e may contain not only sons but also further descendants of e. Both cuts $\Delta(V_e^r)$ and $\Delta(V') = S_e$ are tight for $(\kappa^r)^T y \geq l_r$. Furthermore, all edges in S_e do not have the reduced knapsack weight. Thus by case (1) we have

$$b_e = \sum_{f \in S_e} b_e = \sum_{\bar{f} \in S_e} \gamma_B \kappa_f^r = \gamma_B l_r = \gamma_B \kappa_e^r.$$

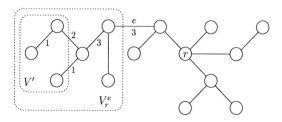

Figure 4.15: Sets V_r and V' considered in case (3b) for $l_\tau = 3$.

Summing up cases (3a) and (3b) we obtain

$$\forall e \in E_B \setminus E_\Pi : \quad b_e = \gamma_B \kappa_e^r.$$

(4) For edges not in E_Π we apply (3). Since all edges in E_Π have the reduced knapsack
weight, we apply the same method as in the case (3b).   □

In the above lemma we already showed the relation between the coefficients of $(\kappa^r)^T y \geq$
l_τ and $b^T y \geq b_0$ for a part of edges in T. The coefficients of the remaining edges are
treated in the following propositions. As we already mentioned above we distinguish two
main cases, (A) and (B), depending on knapsack weights of edges on branch-less paths
to the root.

**(A) At least one maximal branch-less path to the root contains only edges
with the reduced knapsack weight**

Proposition 4.40 *Let the assumptions of Theorem 4.12 (a) hold. Furthermore, assume
that T contains at least one maximal branch-less path to root r having only edges with
the reduced knapsack weight. (4.29) holds if r is a minimal root of T.*

Proof. Let $B_1 = (V_1, E_1), \ldots, B_k = (V_k, E_k)$, $k = \deg(r)$, be all branches in T incident
on r and let $\Pi_i = (V_{\Pi_i}, E_{\Pi_i})$ be the maximal branch-less path to r in B_i, $i = 1, \ldots, k$.
For each branch B_i whose path Π_i contains only edges with the reduced knapsack weight

$$\exists \gamma_i \neq 0 \quad \forall e \in E_i : \qquad b_e = \gamma_i \kappa_e^r. \tag{4.33}$$

due to Lemma 4.39 (4). We distinguish two cases:

(1) Both paths Π_i and Π_j for some i, j, $1 \leq i, j \leq k$, have only edges with the reduced
knapsack weight.

In this case (4.33) holds for $e \in E_i$ and $e \in E_j$, respectively. Since each branch contains at least one edge with the reduced knapsack weight, we take an edge $e_i \in E_i$ and an edge $e_j \in E_j$ with the reduced knapsack weight and apply Lemma 4.39 (1). We have

$$\gamma_i l_r = \gamma_i \kappa^r_{e_i} = b_{e_i} = b_{e_j} = \gamma_j \kappa^r_{e_j} = \gamma_j l_r.$$

Hence there holds

$$\gamma_i = \gamma_j =: \gamma. \tag{4.34}$$

This means, for each edge e in any branch in T whose maximal branch-less path to r has only edges with the reduced knapsack weight we have $b_e = \gamma \kappa^r_e$.

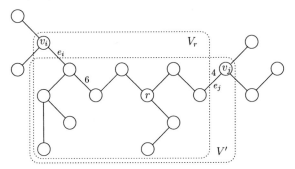

Figure 4.16: Swapping v_i for v_j, $l_r = 6$.

(2) Π_i is a path having all edges with the reduced knapsack weight and Π_j has at least one edge with a non-reduced knapsack weight for some i, j with $1 \leq i, j \leq k$.

In this case

$$\forall\, e \in E_i : \qquad b_e = \gamma \kappa^r_e \tag{4.35}$$

holds by case (1). First we derive an auxiliary relation (4.36). We take an edge $e_j \in E_j$ such that $\kappa^r_{e_j} < l_r$. Using Proposition 4.15 (4) we construct a subset D of non-related edges in E_i such that

$$\sum_{e \in D} \kappa^r_e = l_r - \kappa^r_{e_j}.$$

Hence the cut $\Delta = \{e_j\} \cup D$ is double-tight for $(\kappa^r)^T y \geq l_r$. Let $(V_r, \bar{V}_r)$ be the corresponding bisection. Note that $r \notin \partial V_r \cap V_i$ since all edges in D have non-reduced knapsack weights. We swap a node $v_i \in \partial V_r \cap V_i$ for the end-node of e_j, say v_j, which is not in V_r and obtain the set V', see Figure 4.16. Let e_i be the edge in $\Delta(V')$ adjacent to v_i. If e_i is not a leaf then D contains all sons of e_i. As we already mentioned, $v_i \neq r$ and thus we apply Corollary 4.31 to the cuts $\Delta(V_r) = \{e_j\} \cup S_i \cup \bar{D}$ and $\Delta(V') = \{e_i\} \cup S_j$. S_i and S_j are possibly empty sets of sons of e_i and e_j respectively. By (4.25), (4.20), and (4.35) we have

$$b_{e_j} - \sum_{e \in S_j} b_e = b_{e_i} - \sum_{e \in S_i} b_e = \gamma \Big(\kappa^r_{e_i} - \sum_{e \in S_j} \kappa^r_e \Big) = \gamma. \tag{4.36}$$

Next, we show $b_e = \gamma \kappa_e^r$ for any $e \in B_i$ by the following case distinction concerning the position of the edges in B_j.

(2a) e_j does not lie on Π_j.
By Lemma 4.39 (3) we have

$$\exists \gamma_j \neq 0 \quad \forall e \in E_j \setminus E_{\Pi_j} : \qquad b_e = \gamma_j \kappa_e^r. \tag{4.37}$$

We apply (4.20) and (4.37) for e_j and $e \in S_j$ to (4.36). Thus we obtain

$$\gamma_j = \gamma_j(\kappa_{e_j}^r - \sum_{e \in S_j} \kappa_e^r) = b_{e_j} - \sum_{e \in S_j} b_e = b_{e_i} - \sum_{e \in S_i} b_e = \gamma,$$

i.e.,

$$\forall e \in E_j \setminus E_{\Pi_j} : \qquad b_e = \gamma \kappa_e^r. \tag{4.38}$$

(2b) e_j is the deepest edge in Π_j, i.e., e_j is the leaf in Π_j non-incident on root r.
If $B_j = \Pi_j$ then e_j is a leaf and by (4.36) $b_j = \gamma = \gamma \kappa_{e_j}^r$. If $\kappa_{e_j}^r < l_\tau$ we apply
(4.38) for $e \in S_j \subseteq E_j \setminus E_{\Pi_j}$ and by (4.36) we obtain

$$b_{e_j} = \gamma + \sum_{e \in S_j} b_e = \gamma(1 + \sum_{e \in S_j} \kappa_e^r) = \gamma \kappa_e^r. \tag{4.39}$$

(2c) e_j is an edge in Π_j with $\kappa_{e_j}^r < l_\tau$.
Using (4.39) and recursively applying (4.36) for the edges succeeding the deepest edge in Π_j we derive $b_{e_j} = \gamma \kappa_e^r$ as well.

(2d) e_j is an edge in Π_j with $\kappa_{e_j}^r = l_\tau$.
We apply Lemma 4.39 (1) in the usual manner to the edge e_j and some $e_i \in \Pi_i$
with the reduced knapsack weight and get

$$\gamma_j l_\tau = \gamma_j \kappa_{e_j}^r = b_{e_j} = b_{e_i} = \gamma \kappa_{e_i}^r = \gamma l_\tau.$$

From the above case discussion we conclude

$$\forall e \in E_j : \qquad b_e = \gamma \kappa_e^r.$$

This yields $b_e = \gamma \kappa_e^r$ for each edge e in any branch containing a branch-less path to root having edges with non-reduced knapsack weights.

Summing up the case (1) and (2) we obtain

$$\forall e \in E : \qquad b_e = \gamma \kappa_e^r,$$

i.e., (4.29) holds under the assumptions of Proposition 4.40 so far.   □

(B) All branch-less paths to the root contain edges with non-reduced knap-sack weights

Proposition 4.41 *Let the assumptions of Theorem 4.12 (a) hold. Furthermore, let T be rooted at r such that $\deg(r) = 2$ and assume that both branch-less paths to r contain edges with non-reduced knapsack weights. (4.29) holds if r is minimal and each branch-less path in T contains less than u_τ nodes.*

Proof. Since $\deg(r) = 2$, root r lies on a branch-less path in T. Let $\Pi = (V_\Pi, E_\Pi)$ be the maximal branch-less path in T containing r. Note that $|E_\Pi| \geq 2$. Let $B_1 = (V_1, E_1)$ and $B_2 = (V_2, E_2)$ be the two branches in T incident on the root.

First we show that the assumptions of Proposition 4.41 ensure that none of the end-edges in Π is a leaf in T. Since we assume that $l_\tau > 1$, only one end-edge in Π could be a leaf in T. Hence, w.l.o.g., we assume that $E_1 \setminus E_\Pi \neq \emptyset$ and $E_2 \setminus E_\Pi = \emptyset$. Let g be the deepest edge in $E_1 \cap E_\Pi$ and v the end-node of Π incident to g. Since κ_g^r is not reduced, we have $|V_g^r| \leq l_\tau$ and thus

$$u_\tau \leq |V \setminus V_g^r| = |V_\Pi \setminus \{v\}| < |V_\Pi|$$

contradicting the assumption that each branch-less path in T has less than u_τ nodes. Therefore we obtain

$$E_1 \setminus E_\Pi \neq \emptyset \quad \text{and} \quad E_2 \setminus E_\Pi \neq \emptyset. \tag{4.40}$$

Furthermore, by Lemma 4.39 (3) we have for $i = 1, 2$

$$\exists \gamma_i \neq 0 \quad \forall e \in E_i \setminus E_{\Pi_i} : \qquad b_e = \gamma_i \kappa_e^r. \tag{4.41}$$

We distinguish the following two cases.

(1) Edges in $E \setminus E_\Pi$.
Let g be the deepest edge in $E_1 \cap E_\Pi$ and S_g be the set of its sons. By (4.40) we have $S_g \neq \emptyset$. By the assumption g does not have the reduced knapsack weight, hence by (4.20)

$$\sum_{e \in S_g} \kappa_e^r = \kappa_g^r - 1 < |V_g^r| \leq l_\tau.$$

Using Corollary 4.27 and the assumption that $|V_\Pi| < u_\tau$ we find a bisection $(V_r, \bar{V}_r)$ of V such that

(a) $S_g \subset \Delta(V_r)$,

(b) $V_\Pi \subset V_r$,

(c) $|V_r| = u_\tau$ and the subgraph $(V_r, E(V_r))$ of T is connected,

(d) $V_r \cap (V_2 \setminus V_\Pi) \neq \emptyset$,

(e) $V_2 \cap \bar{V}_r \neq \emptyset$,

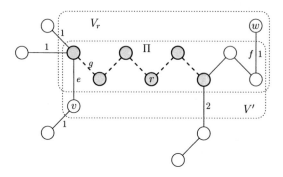

Figure 4.17: Path Π and node sets V_r, V' considered in case (1) for $l_r = 6$.

see Figure 4.17. Due to (c) and Lemma 4.19 the bisection $(V_r, \bar{V}_r)$ is double-tight for $(\kappa^r)^t y \geq l_r$. Thus we apply Lemma 4.29. We swap a node $w \in \partial V_r \cap V_2$ for a node v adjacent to one of the edges in S_g and obtain the set V', see again Figure 4.17. By (d) $w \notin V_\Pi$, thus in particular $w \neq r$. Hence we apply Corollary 4.31 to the corresponding bisection cuts

$$\Delta(V_r) = \{e\} \cup S_f \cup D \quad \text{and} \quad \Delta(V') = S_e \cup \{f\} \cup D,$$

where e is the edge in S_g incident on v, S_e is the set of sons of e, $f \in E_2 \setminus E_\Pi$ is the edge joining w and a node in V', S_f is the set of its sons and D is the set of other edges in both cuts related neither to e nor f. By (4.25), (4.20) and (4.41) we have

$$\gamma_2 = \gamma_2(\kappa^r_f - \sum_{\bar{e} \in S_f} \kappa^r_{\bar{e}}) = b_f - \sum_{\bar{e} \in S_f} b_{\bar{e}} = b_e - \sum_{\bar{e} \in S_e} b_{\bar{e}} = \gamma_1(\kappa^r_e - \sum_{\bar{e} \in S_e} \kappa^r_{\bar{e}}) = \gamma_1 =: \gamma.$$

This, again by (4.41), yields

$$\forall e \in E \setminus E_\Pi : \qquad b_e = \gamma \kappa^r_e. \tag{4.42}$$

(2) Edges in E_Π.
Here we consider only edges in $E_\Pi \cap E_1$. Edges in $E_\Pi \cap E_2$ can be handled in an analogous way. Let $(V_r, \bar{V}_r)$ be the bisection considered in case (1). This time we construct V' by swapping the end-node u of g, which is simultaneously the end-node of Π in B_1, for some node $s \in V_2 \cap \bar{V}_r$ adjacent to a node in V_r, due to (e). See also Figure 4.18. Since u is the end-node of a maximal branch-less path, we have $\deg(u) \neq 2$, i.e., $u \neq r$. We apply as usual Corollary 4.31 to the cuts

$$\Delta(V_r) = \{f\} \cup S_g \cup \bar{D} \quad \text{and} \quad \Delta(V') = \{g\} \cup S_f \cup \bar{D},$$

where $f \in E_2 \setminus E_\Pi$ joins s with a node in V_r, S_f is the set of sons of f and $\bar{D}$ is the subset of possible other edges in both cuts.

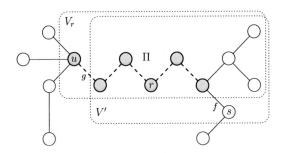

Figure 4.18: Sets V_r and V' considered in case (2), $l_\tau = 6$.

We apply (4.25) to the cuts $\Delta(V_r)$ and $\Delta(V')$, and (4.42) to edges in $S_g \cup S_f \cup \{f\}$, which are in $E \setminus E_\Pi$. We get

$$b_g = \sum_{e \in S_g} b_e + b_f - \sum_{e \in S_f} b_e = \sum_{e \in S_g} \gamma \kappa_e^r + \gamma(\kappa_f^r - \sum_{e \in S_f} \kappa_e^r) = \gamma(\sum_{e \in S_g} \kappa_e^r + 1) = \gamma \kappa_g^r, \quad (4.43)$$

where the last two equalities hold by (4.20).

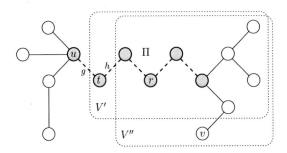

Figure 4.19: Sets V' and V'' considered in case (2a), $l_\tau = 6$.

Next, we consider the other edges in $E_\Pi \cap E_1$. Let h be the father of g.

(2a) h does not have the reduced knapsack weight.

   The bisection $(V', \bar{V}')$ constructed above is double-tight for $(\kappa^r)^T y \geq l_\tau$ due to Lemma 4.29. Note that $\bar{V}' \cap (V_2 \setminus V_\Pi) \neq \emptyset$. We swap a node adjacent to u, say t, for some node in $\bar{V}'$, say v, adjacent to a node in $\partial V'$, see Figure 4.19. The so constructed set V'' contains r since t is on the path to r. We derive $b_h = \gamma \kappa_h^r$ in the same way as (4.43).

(2b) h has the reduced knapsack weight.

   In this case $\kappa_g^r = l_\tau$ holds. We apply Lemma 4.39 (1) and get

$$b_h = b_g = \gamma l_\tau = \gamma \kappa_h^r.$$

Similarly to h we deal with edges preceding it on the path to r and so we obtain

$$\forall e \in E_\Pi \cap E_1 : \qquad b_e = \gamma \kappa_e^r.$$

Summing up cases (1) and (2) we obtain

$$b_e = \gamma \kappa_e^r, \quad \forall e \in E,$$

i.e., (4.29) holds under assumptions of Proposition 4.41. $\qquad\qquad\qquad\qquad$ □

Proposition 4.42 *Let the assumptions of Theorem 4.12 (a) hold. Furthermore, let $r \in V$ with $\deg(r) = 3$ and assume that all branch-less paths to r contain edges with non-reduced knapsack weights. (4.29) holds if*

(a) *r is minimal and*

(b) *each branch-less path in T contains less than u_r nodes or*

(c) *a branch-less path in T contains exactly u_r nodes and an end-edge of this path is a leaf in T.*

Proof. Let $B_1 = (V_1, E_1), \dots, B_k = (V_k, E_k)$, with $k = \deg(r)$, be all branches in T incident on r and let $\Pi_i = (V_{\Pi_i}, E_{\Pi_i})$ be the maximal branch-less path to root in B_i, $i = 1, \dots, k$. At first we are going to show that for each $i = 1, \dots, k$ there exists some $\gamma_i \neq 0$ such that

$$\forall e \in E_i : \qquad b_e = \gamma_i \kappa_e^r.$$

Afterwards we apply Lemma 4.39 to derive

$$\forall 1 \leq i, j \leq k : \quad \gamma_i = \gamma_j =: \gamma.$$

Let B_i, $1 \leq i \leq k$, be some branch incident on r. Note that if $|E_i| = 1$ then $b_e = \gamma_i \kappa_e^r$ holds trivially for the only edge $e \in E_i$. Hence in the following we assume $|E_i| > 1$. Furthermore, due to Proposition 4.15 (2) the assumption that Π_i contains edges with non-reduced knapsack weights provides that the knapsack weight of the deepest edge in Π_i is not reduced.

In Lemma 4.39 (3) we showed already that

$$\forall e \in E_i \setminus E_{\Pi_i} : \qquad b_e = \gamma_i \kappa_e^r. \tag{4.44}$$

We merely need to show that $b_e = \gamma_i \kappa_e^r$ for all $e \in E_{\Pi_i}$. For that we distinguish two cases depending whether B_i coincides with its branch-less path to root or not. In the course of the proof we will frequently apply a sequence of node swaps specified below.

Let $v, w \in V_i$ and $v, w \neq r$. We construct a bisection $(V_{vw}, \bar{V}_{vw})$ such that $r, v, w \in V_{vw}$, $(V_{vw}, E(V_{vw}))$ is a connected subgraph of T and $|V_{vw}| = u_r$, or equivalently, due to

Lemma 4.19, the bisection cut $\Delta(V_{vw})$ is double-tight for $(\kappa^r)^T y \geq l_r$. Next, we find nodes $s, t \in \bar{V}_{vw} \setminus V_i$ adjacent to nodes in V_{vw} but not to each other. Then we swap v and w consecutively for s and t. Due to our construction none of the nodes v, w, s, t coincides with r, hence we apply Corollary 4.31 pairwise to the corresponding cuts, see e.g. Figure 4.22. Let $e = vw$, f be the father of e, let e_1 join s with a node in V_{vw} and e_2 join t with a node in V_{vw}. Finally, let S_e, S_f, S_1, S_2 be the set of sons of e, f, e_1, e_2 respectively, empty if the corresponding edge is a leaf. Then $\Delta(V_{vw})$ takes the following form

$$\Delta(V_{vw}) = S_e \cup (S_f \setminus \{e\}) \cup \{e_1, e_2\} \cup D,$$

where D is the set of all other possible edges in the cut. By swapping v for s we get

$$\Delta(V_{ws}) = S_f \cup S_1 \cup \{e_2\} \cup D$$

and by swapping s for t we have

$$\Delta(V_{wt}) = S_f \cup S_2 \cup \{e_1\} \cup D.$$

Finally, we swap w for s and obtain

$$\Delta(V_{ts}) = \{f\} \cup S_1 \cup S_2 \cup D.$$

Next, we apply consecutively relation (4.25) to $(\Delta(V_{vw}), \Delta(V_{ws}))$, $(\Delta(V_{ws}), \Delta(V_{wt}))$ and $(\Delta(V_{wt}), \Delta(V_{ts}))$

$$b_e - \sum_{\bar{e} \in S_e} b_{\bar{e}} = b_{e_1} - \sum_{e \in S_1} b_e = b_{e_2} - \sum_{e \in S_2} b_e = b_f - \sum_{e \in S_f} b_e. \tag{4.45}$$

For ease of exposition we will call the above procedure a **triple-swap**. The crucial step to apply a triple-swap is to ensure the existence of the nodes s and t in $\bar{V}_{vw}$. We work this out differently in each case considered below.

(1) $E_i = E_{\Pi_i}$.

Let e be the deepest edge in Π_i and f its father ($|E_i| \geq 2$), see Figure 4.20. $|V_i| \leq u_r$ holds by the assumption of the lemma. Hence the existence of a bisection $(V_{vw}, \bar{V}_{vw})$ such that $V_i \subset V_{vw}$ ($r \in V_{vw}$), the subgraph $(V_{vw}, E(V_{vw}))$ of T is connected, and $|V_{vw}| = u_r$ is ensured. Therefore $\Delta(V_{vw}) \subset E \setminus E_i$ and since $|\bar{V}_{vw}| = l_r \geq 2$ we may choose V_{vw} so that s and t in $\bar{V}_{vw}$ are adjacent to nodes in V_{vw}, see Corollary 4.28. Thus we may apply a triple-swap. Note that since e is a leaf we have $S_e \cup S_f \setminus \{e\} = \emptyset$ and $S_f = \{e\}$. The corresponding cuts take the form

$$\begin{aligned}
\Delta(V_{vw}) &= \{e_1, e_2\} \cup D \\
\Delta(V_{ws}) &= \{e, e_2\} \cup S_1 \cup D, \\
\Delta(V_{wt}) &= \{e, e_1\} \cup S_2 \cup D, \\
\Delta(V_{ts}) &= \{f\} \cup S_1 \cup S_2 \cup D,
\end{aligned}$$

see Figure 4.20.

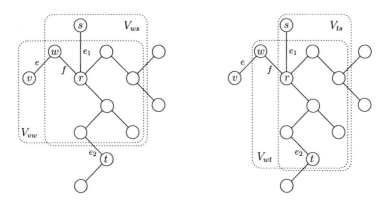

Figure 4.20: The sets V_{vw}, V_{ws}, V_{wt}, and V_{ts} considered in case (1), $l_\tau = 5$.

Applying now (4.45) we obtain

$$\gamma_i = b_e = b_{e_1} - \sum_{\bar{e} \in S_1} b_{\bar{e}} = b_{e_2} - \sum_{\bar{e} \in S_2} b_{\bar{e}} = b_f - b_e. \qquad (4.46)$$

Since $b_e = \gamma_i$, by Lemma 4.39 (2) we have

$$b_f = 2\gamma_i = \gamma_i \kappa_f^r. \qquad (4.47)$$

To determine the coefficients of the edges preceding f on the branch-less path to root r we refine further the case distinction depending on the number of edges in E_i and the value l_τ.

(1a) $|E_i| = 2$.
There is nothing to show anymore, since the two edges are e and f considered above.

(1b) $|E_i| \geq 3$ and $l_\tau = 2$.
In this case all $h \in E_i \setminus \{e, f\}$ have the reduced knapsack weight. Using by default Lemma 4.39 (1) we get

$$b_h = b_f = 2\gamma_i = \gamma_i \kappa_h^r.$$

(1c) $|E_i| \geq 3$ and $l_\tau \geq 3$.
Again we construct an appropriate bisection $(V_{vw}, \bar{V}_{vw})$ to apply a triple-swap. This time we select the edge f from the path E_i so that $2 \leq |V_f^r| \leq l_\tau$, i.e., κ_f^r is not reduced. The edge e is, as assumed in the definition of the triple-swap, the son of f. Let d be the son of e. Note that in the considered case

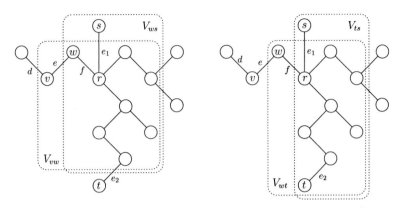

Figure 4.21: The sets V_{vw}, V_{ws}, V_{wt}, and V_{ts} considered in case (1c), $l_\tau = 5$.

there exists at least one such triple of edges $\{f, e, d\}$. We have $|V_d^r| \leq l_\tau - 2$, therefore, $|\bar{V}_{vw} \setminus V_d^r| \geq 2$ and by Corollary 4.28 there are some non-adjacent nodes $s, t \in \bar{V}_{vw}$, $s \neq t$.

The triple-swap yields the following cuts

$$\begin{aligned}
\Delta(V_{vw}) &= \{d, e_1, e_2\} \cup \bar{D}, \\
\Delta(V_{ws}) &= \{e, e_2\} \cup S_1 \cup \bar{D}, \\
\Delta(V_{wt}) &= \{e, e_1\} \cup S_2 \cup \bar{D}, \\
\Delta(V_{ts}) &= \{f\} \cup S_1 \cup S_2 \cup \bar{D},
\end{aligned}$$

see Figure 4.21. By (4.45) and (4.46) we have

$$\gamma_i = b_e - b_d = b_1 - \sum_{e \in S_1} b_e = b_2 - \sum_{e \in S_2} b_e = b_f - b_e. \tag{4.48}$$

Note that in the first triple which we consider we have $|V_f^r| = 3$. Hence by (4.47) and (4.48)

$$b_f = b_e + \gamma_i = 3\gamma_i = \gamma_i \kappa_f^r. \tag{4.49}$$

Next, we exchange recursively f with its father, e with f, and d with e, as long as $|V_f^r| \leq l_\tau$. In this way we obtain $b_f = \gamma_i \kappa_f^r$ from the triple-swap and the corresponding equation (4.48). If f has the reduced knapsack weight then $|V_e^r| = l_\tau$. Now, we apply Lemma 4.39 (1) to edges e, f and each edge h preceding f on the path to r

$$b_h = b_f = b_e = \gamma_i \kappa_e^r = \gamma_i l_\tau = \gamma_i \kappa_f^r.$$

Summing up,

$$b_e = \gamma_i \kappa_e^r$$

holds for all $e \in E_{\Pi_i}$ if B_i coincides with path Π_i.

(2) $E_i \setminus E_{\Pi_i} \neq \emptyset$.

In the considered case we have $|E_i| \geq 3$, since B_i contains at least two leaves and Π_i at least one edge. Furthermore, since the deepest edge in Π_i has a non-reduced knapsack weight this case occurs only if $l_\tau \geq 3$. We construct a triple-

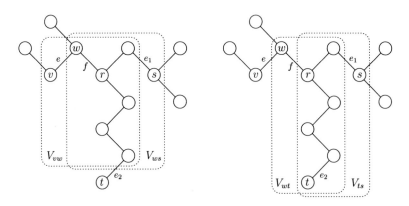

Figure 4.22: The sets V_{vw}, V_{vs}, V_{vt}, and V_{ts} considered in case (2), $l_\tau = 6$.

swap. Let f be the deepest edge of the path Π_i and e one of its sons. Note that due to our assumptions f is neither a leaf nor has the reduced knapsack weight. By Proposition 4.15 (2), (3) the knapsack weights of all descendants of f are not reduced. Hence

$$\sum_{\bar{e} \in S_e} \kappa_{\bar{e}}^r + \sum_{\bar{e} \in S_f \setminus \{e\}} \kappa_{\bar{e}}^r = \sum_{\bar{e} \in S_e} |V_{\bar{e}}^r| + \sum_{\bar{e} \in S_f \setminus \{e\}} |V_{\bar{e}}^r| < |V_f^r| \leq l_\tau. \qquad (4.50)$$

By Corollary 4.27 there exists a bisection cut which contains $S_e \cup S_f \setminus \{e\}$ and is double-tight for $(\kappa^r)^T y \geq l_\tau$. Let Δ be such a cut and $(V_r, \bar{V}_r)$ be the corresponding bisection. Obviously the nodes v, w and r belong to V_r, see Figure 4.22. Hence we set $V_{vw} := V_r$, $\bar{V}_{vw} := \bar{V}_r$ and $\Delta(V_{vw}) := \Delta$. Furthermore, we have

$$|\bar{V}_{vw} \cap V_i| = |V_f^r \setminus \{v, w\}| \leq l_\tau - 2 \ .$$

This together with Corollary 4.28 ensures the existence of two non-adjacent nodes s, t in $\bar{V}_{vw} \setminus V_i$. Thus a triple-swap can be applied. The cuts $\Delta(V_{vw}), \Delta(V_{ws}), \Delta(V_{wt})$, and $\Delta(V_{ts})$ take the form

$$
\begin{aligned}
\Delta(V_{vw}) &= S_e \cup (S_f \setminus \{e\}) \cup \{e_1, e_2\} \cup D, \\
\Delta(V_{ws}) &= S_f \cup S_1 \cup \{e_2\} \cup D, \\
\Delta(V_{wt}) &= S_f \cup S_2 \cup \{e_1\} \cup D, \\
\Delta(V_{ts}) &= \{f\} \cup S_1 \cup S_2 \cup D.
\end{aligned}
$$

By (4.44) we have $b_{\bar{e}} = \gamma_i \kappa_{\bar{e}}^r$ for $\bar{e} \in S_f \cup S_e \cup \{e\}$. Using (4.45) and (4.20) we obtain

$$b_f = b_e - \sum_{\bar{e} \in S_e} b_{\bar{e}} + \sum_{\bar{e} \in S_f} b_{\bar{e}} = \gamma_i (1 + \sum_{\bar{e} \in S_f} \kappa_{\bar{e}}^r) = \gamma_i \kappa_f^r.$$

Analogously to case (1c) we deal with remaining edges on Π_i.

Summing up the above cases (1) and (2) we obtain so far that

$$\forall e_i \in E_i : \qquad b_{e_i} = \gamma_i \kappa_{e_i}^r \qquad (4.51)$$

holds for an arbitrary branch B_i, $1 \leq i \leq k$, incident on root r.

In the above case distinction we showed that if $|E_i| > 1$ for some branch B_i, $1 \leq i \leq k$, then there exists at least one triple-swap in B_i. Let B_j, $j \neq i$, be some other branch in T incident on r and let $(V_{vw}, \bar{V}_{vw})$ be the initial bisection for some triple-swap in B_i. Since $|\bar{V}_{vw} \setminus V_i| \geq 2$, we construct V_{vw} in such a way that $\bar{V}_{vw} \cap V_j \neq \emptyset$ and take s from V_j. By (4.44) we have

$$\gamma_i = b_e - \sum_{\bar{e} \in S_e} = b_{e_1} - \sum_{\bar{e} \in S_1} = \gamma_j.$$

If T is a star, i.e., for each $i = 1, \ldots, k$ we have $|E_i| = 1$, then by the assumption $l_r > 1$ all edges in T do not have reduced knapsack weight. Lemma 4.39 (2) yields

$$\gamma_i = b_{e_i} = b_{e_j} = \gamma_j$$

for $e_i \in E_i$ and $e_j \in E_j$, $1 \leq i, j, \leq k$. Hence $\gamma_i = \gamma_j =: \gamma$ holds for all $1 \leq i, j \leq k$. Finally, using (4.51) we obtain

$$\forall e \in E : \qquad b_e = \gamma \kappa_e^r.$$

This completes the proof of Proposition 4.42. $\square$

In Theorem 4.12 the sufficient and necessary conditions for $\kappa^T y \geq l_r$ to be facet-defining inequality for P_B depend on the value of l_r. Propositions 4.40 - 4.42 treat cases valid for $l_r > 1$, which we summarize in the next corollary. The following lemma concerns the remaining case, $l_r = 1$.

Corollary 4.43 *Let the assumptions of Theorem 4.12 (a) hold. Let F_κ be the face P_B induced by $\kappa^T y \geq l_r$ and F_b be a face of P_B defined by the inequality $b^T y \geq b_0$ such that $F_\kappa \subset F_b$. Assume that*

(a) *T is rooted at a minimal root and*

(b) *either the maximal branch-less path in T has less than u_r nodes or*

(c) *exactly u_r nodes and one of the end-edges of this branch-less path is a leaf in T.*

Then there exists $\gamma > 0$ such that $b_e = \gamma \kappa_e^r$ for all $e \in E$ and $b_0 = \gamma l_\tau$, i.e.,

$$F_\kappa = F_b.$$

Proof. Let the assumptions of Theorem 4.12 (a) hold. Proposition 4.40 treats the case (A), where at least one maximal branch-less path to the root contains only edges with the reduced knapsack weight. Propositions 4.41 and 4.42 handle the complementary case (B), where all branch-less paths to the root contain edges with non-reduced knapsack weights. For these both possible situations there exists some $\gamma \neq 0$ such that $b_e = \gamma \kappa_e^r$ for all $e \in E$. Let Δ be a bisection cut, whose incidence vector χ^Δ lies on the face $F_\kappa \subset F_b$. We have

$$b_0 = \sum_{e \in E} b_e \chi_e^\Delta = \sum_{e \in \Delta} b_e = \gamma \sum_{e \in \Delta} \kappa_e^r = \gamma l_\tau.$$

Hence $\kappa^T y \geq l_\tau$ coincides with $b^T y \geq b_0$ and thus $F_\kappa = F_b$. □

If $l_\tau = 1$ we obtain the following result.

Lemma 4.44 *Let the assumptions of Theorem 4.12 (b) hold. Then there are $|E|$ bisection cuts tight for $(\kappa^r)^T y \geq l_\tau$, whose incidence vectors are linearly independent.*

Proof. If $l_\tau = 1$ then for all $e \in E$ we have $\kappa_e^r = 1$ for any selection of root r in V. The knapsack tree inequality takes the form

$$\sum_{e \in E} y_e = 1.$$

For each $e \in E$ the cut $\Delta = \{e\}$ is a feasible bisection cut tight for the knapsack inequality. Thus the claim of the lemma follows. □

The intermediate steps stated in Propositions 4.40 - 4.42, Corollary 4.43 and Lemma 4.44 contribute to the proof of the main theorem which can now be finalized.

Proof of Theorem 4.12. Let F_κ be the face $P_\mathcal{B}$ induced by $\kappa^T y \geq l_\tau$ and F_b be a face of $P_\mathcal{B}$ defined by the inequality $b^T y \geq b_0$ such that $F_\kappa \subset F_b$. Under assumptions of Theorem 4.12 on page 59 the polytope $P_\mathcal{B}$ is full-dimensional. We distinguish two cases:

(1) $l_\tau > 1$.

Due to Theorem 4.7 the assumption on r being a minimal root in T is necessary for $(\kappa^r)^T y \geq l_\tau$ to be a facet defining inequality for $P_\mathcal{B}$.

By Corollary 4.38 it is necessary that each branch-less path in T contains less than u_τ nodes or exactly u_τ nodes and one of its end-edge is a leaf in T.

Let F_b be a facet of $P_\mathcal{B}$ defined by inequality $b^T y \geq b$ and such that $F_\kappa \subseteq F_b$. By Corollary 4.43 there exists $\gamma \neq 0$ such that

$$b_e = \gamma \kappa_e^r, \quad \forall e \in E,$$

$$b_0 = \gamma l_\tau.$$

Due to Proposition A.1 F_κ is a facet of $P_\mathcal{B}$.

(2) $l_\tau = 1$.

By Lemma 4.44 $\dim(F_\kappa) = |E| - 1$ and the claim follows from the definition of a facet. $\qquad\square$

4.2.6 Minimal Root as Sufficient and Necessary Condition

The case considered in Proposition 4.40 is independent from the node length of the branch-less paths in T and the degree of root r. Namely, the assumption on r to be minimal forces that any branch-less path in T cannot have too many nodes, as we show in the next lemma.

Lemma 4.45 *Let T be rooted at $r \in V$ and assume that at least one branch-less path to r contains only edges with the reduced knapsack weight. If r is minimal then either each branch-less path in T contains less than u_τ nodes or one end-edge of a maximal branch-less path in T with exactly u_τ nodes is a leaf in T.*

Proof. Let $B = (V_B, E_B)$ be a branch in T, whose maximal branch-less path to root $\Pi' = (V', E')$ contains only edges with the reduced knapsack weight. Denote by s the end-node of Π' not equal to r and by h the end-edge of Π' incident on s. Let $\Pi = (V_\Pi, E_\Pi)$ be a branch-less path in T such that $|V_\Pi| \geq u_\tau$. Denote by $v, w \in V_\Pi$ the end-nodes and $e, f \in E_\Pi$ the end-edges, incident on v and w respectively. Note that due to the assumption that r is a minimal root of T we have

(A) $r \in V_\Pi$

by Proposition 4.32 (3) and

(B) $|V_h^r| > l_\tau$

follows from the assumption that all edges on the path Π' ha ve the reduced knapsack weight. We consider several cases depending on the common component of the paths Π' and Π:

(1) $V' \subseteq V_\Pi$, i.e., Π' is a sub-path of Π.

Since Π' is a maximal branch-less path in B, end-node s must be the end-node of Π, say $s = v$.

(1a) $|V_\Pi| \geq u_\tau + 1$.

By (B) we obtain the contradiction

$$|V| \geq |V_h^r \setminus \{s\}| + |V_\Pi| \geq l_\tau + u_\tau + 1 > |V|.$$

(1b) $|V_\Pi| = u_\tau$ and none of the end-edges of Π is a leaf in T.

In this case $\deg(w) \geq 3$, since w is an end-node of a maximal branch-less path and not a leaf. Therefore $|V_f^r| \geq 3$ and (B) yield the following contradiction

$$|V| = |V_h^r \setminus \{s\}| + |V_\Pi| + |V_f^r \setminus \{w\}| \geq l_\tau + u_\tau + 2 > |V|.$$

(2) $V' \setminus V_\Pi \neq \emptyset$ and $|V' \cap V_\Pi| \geq 2$.

In this case the paths Π and Π' compose together a branch-less path, say $\bar{\Pi}$. Since Π' is a maximal branch-less path in B, end-node s must be the end-node of $\bar{\Pi}$ and root r must be some inner node of $\bar{\Pi}$. By (B) we obtain the contradiction

$$|V| \geq |V_h^r \setminus \{s\}| + |V' \setminus V_\Pi| + |V_\Pi| \geq l_r + 1 + u_r > |V|.$$

(3) $|V' \cap V_\Pi| = 1$.

$V' \cap V_\Pi = \{s\}$ is not possible due to (A). Hence we assume that $r = v$. By (B) we obtain the contradiction

$$|V| \geq |V_h^r| + |V_\Pi| > l_r + u_r = |V|.$$

Note that due to (A) Π and Π' share root node r. Hence the case $V' \cap V_\Pi = \emptyset$ is not possible.   □

Due to the above lemma we can point out a special case, when only a minimal root r of T suffices for $(\kappa^r)^T y \geq l_r$ to define a facet of the bisection cut polytope.

Theorem 4.46 *Let the assumptions of Theorem 4.12 hold. Furthermore, assume that at least one maximal branch-less path to r in T contains only edges with reduced knapsack weights. $(\kappa^r)^T y \geq l_r$ is a facet-defining inequality for $P_\mathcal{B}$ if and only if r is a minimal root.*

4.2.7 Equipartition

In the exact equipartition case $l_r = u_r$, which occurs only if $|V|$ is an even number, the polytope $P_\mathcal{B}$ is full-dimensional if and only if T is neither a star nor a path with node length less than 4, due to Corollary 2.13. Because of these restrictions we consider the equipartition case separately.

Theorem 4.47 *Given a tree $T = (V, E)$ which is not a star, assume that $|V| \geq 4$, $f_i = 1$ for all $i \in V$, and $l_r = u_r$.*

(1) *If $P_\mathcal{B}$ is full-dimensional then the knapsack tree inequality $(\kappa^r)^T y \geq l_r$ defines a facet for $P_\mathcal{B}$ if and only if T is rooted at a minimal root and each branch-less path in T contains less than u_r nodes or exactly u_r nodes and one end-edge of this path is a leaf in T.*

(2) *If T is a path of node length equal to 4 then $(\kappa^r)^T y \geq l_r$ defines a facet for $P_\mathcal{B}$.*

Proof. We distinguish two cases depending on the dimension of the polytope $P_\mathcal{B}$.

(1) $P_{\mathcal{B}}$ is full-dimensional.

First we show that all edges in T have non-reduced knapsack weights if the root of T is minimal. We use the obvious fact that if T contains edges with the reduced knapsack weight then at least one such edge is incident on root r. Assume that $e = sr \in E$, $s \in V$, has the reduced knapsack weight. Then we have $|V_e^r| \geq l_\tau + 1$ and

$$|V_e^s| = |V| - |V_e^r| \leq u_\tau + l_\tau - (l_\tau + 1) = l_\tau - 1.$$

Hence r cannot be a minimal root by Proposition 4.9.

The claim of the theorem follows from Propositions 4.41 and 4.42. (Compare the proof of Theorem 4.12.)

(2) By Corollary 2.13 $P_{\mathcal{B}}$ is not full-dimensional. In this case $P_{\mathcal{B}}$ is the convex hull of the points

$$\begin{pmatrix} 0 \\ 1 \\ 0 \end{pmatrix}, \quad \begin{pmatrix} 1 \\ 0 \\ 1 \end{pmatrix}, \quad \begin{pmatrix} 1 \\ 1 \\ 1 \end{pmatrix},$$

see Figure 4.23. Hence $P_{\mathcal{B}}$ lies on the hyperplane $\{y \in \mathbb{R}^3 \; : \; y_1 = y_3\}$ and

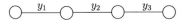

Figure 4.23: The path of node length 4.

$\dim(P_{\mathcal{B}}) = 2$. The knapsack tree inequality

$$y_1 + 2y_2 + y_3 \geq 2$$

reduces to

$$y_1 + y_2 \geq 1$$

and induces a face of $P_{\mathcal{B}}$ of the dimension 1, i.e., it is facet-defining inequality for $P_{\mathcal{B}}$. $\qquad\square$

Remark 4.48 *If T is a star and $l_\tau = u_\tau$ then $P_{\mathcal{B}}$ has the dimension $|E| - 1$. In this case the bisection cut polytope is a subset of the hyperplane induced by the equality*

$$\sum_{e \in E} y_e = l_\tau$$

coinciding with the knapsack tree (or star) inequality.

4.2.8   Min-Cut

If all nodes are weighted with 1 and $l_\tau = 1$ then the MGBP reduces to the min-cut problem on undirected graphs, which is defined as follows. Given an undirected graph $G = (V, E)$ with non-negativ weighted edges. For nodes $s, t \in V$ an **s-t cut** in G is a cut $\Delta(S)$, $S \subset V$, such that $s \in S$ and $t \in V \setminus S$. Finding the cut of a minimum weight among all s-t cuts in G is the **min-cut** problem for undirected graphs.

Following Theorem 4.12 (b) the knapsack tree inequality $(\kappa^r)^T \geq l_\tau$ defines a facet of the bisection cut polytope $P_\mathcal{B}$ for $l_\tau = 1$, i.e., in the min-cut case. It turns out that it is the only nontrivial inequality in the full description of $P_\mathcal{B}$.

Proposition 4.49 *If $l_\tau = 1$, the graph G is a tree with $|E| \geq 3$ and $f_i = 1$, $i \in V$ then $P_\mathcal{B}$ is fully described by the following inequalities*

$$\sum_{e \in E} y_e \;\geq\; 1, \tag{4.52}$$

$$y_e \;\geq\; 0, \quad \forall e \in E, \tag{4.53}$$

$$y_e \;\leq\; 1, \quad \forall e \in E. \tag{4.54}$$

All these inequalities define facets of $P_\mathcal{B}$.

Proof. In the considered case $P_\mathcal{B}$ is defined by the convex hull of all 0-1 vectors in $\mathbb{R}^{|E|} \setminus \{0\}$. Hence inequalities (4.52) - (4.54) are valid for $P_\mathcal{B}$. It is easy to verify that the constraints (4.52) - (4.54) lead to a totally unimodular matrix. Using Proposition A.10 we obtain that the polyhedron

$$P := \{x \in \mathbb{R}^{|E|} \,:\, (4.52), (4.53), \text{ and } (4.54)\}$$

describes the bisection cut polytope. It remains to show that each of inequalities (4.52) - (4.54) is necessary in the description of $P_\mathcal{B}$. Consider the face

$$F := \{y \in P_\mathcal{B} \,:\, \sum_{e \in E} y_e = 1\}$$

defined by the inequality (4.52). Each unit vector in $\mathbb{R}^{|E|}$ lies on F. Thus $\dim(F) = |E| - 1$ and F defines a facet of $P_\mathcal{B}$. Next, we consider the face $F_0 := \{y \in P_\mathcal{B} \,:\, y_e = 0\}$ for some $e \in E$. The $m := |E| - 1$ vectors y_i, $i = 1, \ldots, m$, defined as

$$y_i^k = \begin{cases} 1, & \text{if } i = k \text{ and } i \neq e, \\ 0, & \text{otherwise,} \end{cases} \quad i, k = 1, \ldots, m,$$

and the vector y^0 with

$$y_i^0 = \begin{cases} 0, & \text{if } i = e, \\ 1, & \text{otherwise,} \end{cases}$$

lie on F_0. It is easy to verify that

$$\begin{pmatrix} y^0 \\ 1 \end{pmatrix}, \begin{pmatrix} y^1 \\ 1 \end{pmatrix}, \ldots, \begin{pmatrix} y^m \\ 1 \end{pmatrix} \in \mathbb{R}^{|E|+1}$$

are linearly independent. Hence the vectors $y^0, \ldots, y^m$ are affinely independent and thus $\dim(F_0) = m = |E| - 1$, i.e., all inequalities (4.53) are facet-defining for $P_{\mathcal{B}}$. Finally, for some $e \in E$ we consider the vectors $y^1, \ldots, y^m$, $m := |E|$, lying on the face $F_1 := \{y \in P_{\mathcal{B}} : y_e = 1\}$ and defined as

$$y_i^k = \begin{cases} 1, & \text{if } i = k \text{ and } i = e, \\ 0, & \text{otherwise,} \end{cases} \quad i, k = 1, \ldots, m.$$

It is easy to check that the vectors

$$\begin{pmatrix} y^1 \\ 1 \end{pmatrix}, \ldots, \begin{pmatrix} y^m \\ 1 \end{pmatrix} \in \mathbb{R}^{|E|+1}$$

are linearly independent. Hence the vectors $y^1, \ldots, y^m$ are affinely independent and thus $\dim(F_1) = m - 1 = |E| - 1$, i.e., all inequalities (4.54) are facet-defining for $P_{\mathcal{B}}$. $\qquad \square$

Remark 4.50 *If $l_r = 1$ and $|E| = 2$ then $P_{\mathcal{B}}$ is the convex hull of the points*

$$\begin{pmatrix} 1 \\ 1 \end{pmatrix}, \quad \begin{pmatrix} 1 \\ 0 \end{pmatrix}, \quad \begin{pmatrix} 0 \\ 1 \end{pmatrix}.$$

Hence $P_{\mathcal{B}}$ is full-dimensional. In this case only inequalities (4.52) and (4.54) define facets of $P_{\mathcal{B}}$ and fully describe the bisection cut polytope.

4.3 Separation

The exact separation of the knapsack tree inequalities requires a determination of all valid inequalities for the associated knapsack polytope $P_{\mathcal{K}}$. Since the separation problem over $P_{\mathcal{K}}$ is NP-hard, see Remark A.22, the separation problem for the knapsack tree inequalities inherits the complexity.

In Algorithm 4.1 we summarize the separation procedure for a given knapsack inequality

$$\sum_{i \in V} a_i x_i \leq a_0$$

with $a_i \geq 0$ for all $i \in V$. We follow the strategy presented by Ferreira et al. [32] adjusted with the strengthening methods given in Section 4.1.1 - 4.1.3. We determine first the knapsack tree (Step 3). The root r of the tree is selected randomly from the set V. We apply condition (4.10) while looking for a shortest path tree $T = (V_T, E_T)$ rooted at r with respect to the values of edge variables being the current solution of a relaxation $\bar{y}$.

If the corresponding knapsack tree inequality

$$\sum_{e \in E_T} \alpha_e^r y_e \geq \alpha_0$$

is violated, we strengthen it by looking for a minimal root. In Steps 6 - 13 we follow the procedure described in Section 4.1.3 based on Proposition 4.9 and 4.10.

As valid knapsack inequalities for $P_{\mathcal{K}}$ we consider the inequality describing this polytope

$$\sum_{v \in V} f_v x_v \leq u_\tau$$

and the cover knapsack inequalities (A.11). To obtain valid cover inequalities for $P_{\mathcal{K}}$ we solve a problem of the type (CKI), see Section A.5, modified in the following way.

Algorithm 4.1 Knapsack Tree Inequalities Separation Algorithm

---

Input Knapsack inequality $\sum_{i \in V} a_i x_i \leq a_0$, current solution of a relaxation $\bar{y}$.

1: $success := \text{FALSE}$;

2: select a node $r \in V$ such that $a_r \neq 0$;

3: compute a shortest path tree T rooted at r with respect to the edge weights $\bar{y}$; abandon nodes i with $\left(1 - \sum_{e \in E_{ri}} \bar{y}_e\right) \leq 0$;

4: **if** $\sum_{e \in E_T} \alpha_e^r \bar{y}_e \leq \alpha_0$ **then**

5: 　　$success := \text{TRUE}$;

6: 　**repeat**

7: 　　　$strengthen := \text{FALSE}$;

8: 　　　find s such that $\alpha_{rs} = \max\{\alpha_{rv}^r : rv \in E_T\}$;

9: 　　　**if** $\alpha_e^s < \alpha_e^r$ **then**

10: 　　　　$r := s$;

11: 　　　　$strengthen := \text{TRUE}$;

12: 　　**end if**

13: 　**until** $strengthen = \text{FALSE}$

14: **end if**

Output If $success = \text{TRUE}$, inequality $\sum_{e \in E_T} \alpha_e^r y_e \geq \alpha_0$ such that $r \in \mathcal{R}$.

Since we do not always dispose of node variables in the formulation of the MGBP, we have to establish the values of the vector $\bar{x}$ in another way. Here we use the coincidence between the x-variables in a knapsack inequality

$$\sum_{v \in V} a_v x_v \leq a_0,$$

and the terms $\left(1 - \sum_{e \in E_{rv}} y_e\right)$, $v \in V_T$, in the corresponding knapsack tree inequality

$$\sum_{v \in V_T} a_v \left(1 - \sum_{e \in E_{rv}} y_e\right) \leq a_0.$$

In this way we obtain only $\bar{x}$-values for the nodes in the shortest path tree $T = (V_T, E_T)$ computed in Step 3 in Algorithm 4.1. We set $\bar{x}_r := 1$ and

$$\bar{x}_v = \left(1 - \sum_{e \in E_{rv}} \bar{y}_e\right)$$

for all other $v \in V_T$. Note that since each node $i \in V$ with $\left(1 - \sum_{e \in E_{ri}} \bar{y}_e\right) \leq 0$ is excluded from the tree and $\bar{y} \in [0,1]^{|E|}$ we have $\bar{x}_v \in [0,1]$ for all $v \in V_T$. Hence, to obtain a valid cover inequality we solve the problem

$$\min \sum_{v \in V_T} (1 - \bar{x}_v) z_v,$$

(CKI) $\qquad$ s.t. $\quad \sum_{v \in V_T} f_v z_v > u_T,$

$$z \in \{0,1\}^{|V|}.$$

If we change the meaning of z-variables, so that $z_i = 0$ if item i is in the cover and $z_i = 1$ otherwise, then we obtain an equivalent formulation of (CKI) as a knapsack problem,

$$\max \sum_{v \in V_T} (1 - \bar{x}_v) z_v,$$

(CKI2) $\qquad$ s.t. $\quad \sum_{v \in V_T} f_v z_v < \sum_{v \in V_T} f_v - u_T,$

$$z \in \{0,1\}^{|V_T|}.$$

We solve the latter problem using methods from dynamic programming implemented in SCIP, see e.g. [88].

Running first numerical test we observed that the search for cover inequalities slows down considerably the separation routine for the knapsack tree inequalities. However, the cuts based on the cover inequalities are of a good quality. We established empirically that calling the cover separator every third time is a good compromise.

Chapter 5

Bisection Knapsack Path Inequalities

The idea behind the knapsack tree inequality is to express the knapsack condition for nodes in the same cluster as the root of the supporting tree in terms of edge variables. For example, such a cluster consists of nodes for which all edges on the path to the root are not cut, as exploited in the knapsack tree inequality. In the new class of inequalities, called **bisection knapsack path inequalities**, we exploit the fact that in the bisection case two nodes are also in one cluster if the path joining them is cut an even number of times. These inequalities subsume the knapsack tree inequalities and yield more computational flexibility in finding strong valid inequalities for P_B.

Supports of both knapsack tree and bisection knapsack path inequalities are connected subgraphs. Exploiting the weights of the nodes not belonging to the tree one can reduce the capacity of the corresponding knapsack inequality. Such a strengthening extends the support to non-connected substructures and results in the class of the so called **capacity improved bisection knapsack path inequalities**. These stronger conditions yield nonlinear right-hand sides. Considering the convex envelope of this nonlinear function one can show that the supporting hyperplanes are in one-to-one correspondence to the faces of a certain polytope called **cluster weight polytope**. The complete description of the cluster weight polytope delivers the tightest strengthening possible for the capacity reduced bisection knapsack tree inequalities.

The main contribution to this topic was made by Armbruster [4]. The complete description of the cluster weight polytope for the case of a star without restricted capacity is derived there. After giving some necessary definitions we present an alternative, and shorter, way to show that each point of the cluster weight polytope is in one-to-one correspondence to some point from the polytope described by inequalities exhibited in [4]. To establish the coefficients of these inequalities we solve a continuous knapsack problem. We refer also to Armbruster, F., Martin, and Helmberg [7].

5.1   Definition

The special structure of the graph bisection problem implies that whenever there is a path between two nodes of the graph then the two end-nodes of the path have to be in the same cluster if the path is cut an even number of times, and the two end-nodes are in different clusters if the path is cut an odd number of times. Using this fact we take a closer look at the indicator term

$$\omega_{rv} := 1 - \sum_{e \in E_{rv} \setminus H_v} y_e - \sum_{e \in H_v} (1 - y_e),$$

where $\Pi_{rv} = (V_{rv}, E_{rv})$ is the path joining some nodes $r, v \in V$ and $H_v \subseteq E_{rv}$. Observe first that since $y \in [0,1]^{|E|}$ we have

$$\omega_{rv} \leq 1$$

and the value 1 is obtained if and only if

$$H_v = \{e \in E_{rv} : y_e = 1\}.$$

The even cardinality of H_v forces r and v to be in the same cluster and r, v are in different clusters as long $|H_v|$ is odd. If, however, $H_v \neq \{e \in E_{rv} : y_e = 1\}$ then $\omega_{rv} \leq 0$. In this way we obtain new classes of valid inequalities for the polytope P_B.

Definition 5.1 (bisection knapsack path inequalities) *Let $\sum_{v \in V} a_v z_v \leq a_0$, with $a_v \geq 0$ for all $v \in V$, be a valid inequality for the knapsack polytope P_K. Furthermore, let a subset $V' \subseteq V$, a fixed root node $r \in V'$, paths $\Pi_{rv} = (V_{rv}, E_{rv})$ for each $v \in V' \setminus \{r\}$, and sets $H_v \subseteq E_{rv}$ be given.*

If $|H_v|$ is even for each $v \in V' \setminus \{r\}$ then the inequality

$$\sum_{v \in V'} a_v \left(1 - \sum_{e \in E_{rv} \setminus H_v} y_e - \sum_{e \in H_v} (1 - y_e) \right) \leq a_0 \tag{5.1}$$

is called **even bisection knapsack path inequality**.

If $|H_v|$ is odd for each $v \in V' \setminus \{r\}$ then the inequality

$$\sum_{v \in V'} a_v \left(1 - \sum_{e \in E_{rv} \setminus H_v} y_e - \sum_{e \in H_v} (1 - y_e) \right) \leq a_r + a_0 \tag{5.2}$$

is called **odd bisection knapsack path inequality**.

Remark 5.2 *Due to the negative sub-term $\sum_{e \in H_v} (1 - y_e)$ the value of ω_{rv} increases the more edges from H_v are in the cut. Therefore one can extend the path Π_{rv} to a walk, see [4].*

Note that the term ω_{rv} corresponds to

$$1 - \sum_{e \in E_{rv}} y_e$$

in the knapsack tree inequality (4.1). Hence knapsack tree inequalities are a special case of the bisection knapsack path inequalities, where the paths E_{rv} form a tree, all nodes on these paths are contained in V' and $H_v = \emptyset$ for all $v \in V' \setminus \{r\}$.

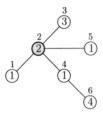

Figure 5.1: Strengthening a knapsack tree inequality to an even bisection knapsack path inequality presented in Example 5.3.

We illustrate on an example how a knapsack tree inequality is strengthened to a even bisection knapsack path inequality.

Example 5.3 Consider the tree depicted in Figure 5.1. The node weights are within the circles and the numbers of the nodes are above the circles. The labeled node corresponds to the root of the tree. For this tree and the knapsack inequality

$$x_1 + 2x_2 + 3x_3 + x_4 + x_5 + 4x_6 \leq 7,$$

the strongest knapsack tree inequality with respect to the considered tree

$$2 + (1 - y_{1,2}) + 3(1 - y_{2,3}) + (1 - y_{2,4}) + (1 - y_{2,5}) + 4(1 - y_{2,4} - y_{4,6}) \leq 7,$$

or equivalently

$$y_{1,2} + 3y_{2,3} + 5y_{2,4} + y_{2,5} + 4y_{4,6} \geq 5,$$

defines a face of P_B of dimension 3. Applying the argument that node 6 will be in the same cluster as the root node 2 if the path joining them is cut an even number of times, we change the term

$$1 - y_{2,4} - y_{4,6}$$

to

$$1 - (1 - y_{2,4}) - (1 - y_{4,6}).$$

Thus we obtain the stronger (even bisection knapsack path) inequality

$$2 + (1 - y_{1,2}) + 3(1 - y_{2,3}) + (1 - y_{2,4}) + (1 - y_{2,5}) + 4(1 - (1 - y_{2,4}) - (1 - y_{4,6})) \leq 7,$$

or in a simpler form

$$y_{1,2} + 3y_{2,3} - 3y_{2,4} + y_{2,5} - 4y_{4,6} \geq -3,$$

defining a 4-dimensional face of $P_\mathcal{B}$.

Proposition 5.4 (Armbruster 2007) *The even bisection knapsack path inequality (5.1) and the odd bisection knapsack path inequality (5.2) are valid for the polytope $P_\mathcal{B}$.*

Proof. We present an alternative proof to that one given in [4]. For ease of exposition we stick to the notation

$$\omega_{vr} := 1 - \sum_{e \in E_{rv} \setminus H_v} y_e - \sum_{e \in H_v} (1 - y_e), \quad v \in V,$$

and follow our observations from the beginning of the section. Let $(S, V \setminus S)$ be a bisection in G with $r \in S$ and $y = \chi^{\Delta(S)}$. Let $\Pi_{rv} = (V_{rv}, E_{rv})$ be a path from r to $v \in V$ in G with $H_v = \{e \in E_{rv} : y_e = 1\}$. If the subsets H_v, $v \in V'$, are all of an even cardinality then

$$\sum_{v \in V'} a_v \omega_{vr} = \sum_{v \in V' \cap S} a_v \omega_{vr} + \sum_{v \in V' \setminus S} a_v \omega_{vr} \leq \sum_{v \in V' \cap S} a_v \leq \sum_{v \in S} a_v \leq a_0.$$

If $|H_v|$ is odd for each $v \in V'$ then

$$\sum_{v \in V'} a_v \omega_{vr} = \sum_{v \in V' \cap S} a_v \omega_{vr} + \sum_{v \in V' \setminus S} a_v \omega_{vr} \leq a_r + \sum_{v \in V' \setminus S} a_v \leq a_r + \sum_{v \in V \setminus S} a_v \leq a_r + a_0.$$

Hence the claim of the proposition follows.   □

Remark 5.5 *In [4] the odd bisection knapsack path inequality is introduced in the form*

$$a_r + \sum_{v \in V' \setminus \{r\}} a_v \left(\sum_{e \in E_{rv} \setminus H_v} y_e + \sum_{e \in H_v} (1 - y_e) \right) \geq a(V') - a_0$$

with odd cardinality sets H_v, $v \in V' \setminus \{r\}$.

5.2   Capacity Reduction Strengthening

To motivate a strengthening for bisection knapsack path inequalities consider the case of a disconnected graph with two components, one of them induced by a single edge uv, the other (connected one) induced by the nodes set $V' = V \setminus \{u, v\}$. Even though one cannot include the edge uv directly in a bisection knapsack path inequality rooted at some $r \in V'$, one can at least tighten the inequality if $y_{uv} = 1$. In this case u and v belong to different clusters and therefore the capacity u_r of both clusters can be reduced by $\min\{f_u, f_v\}$. Similarly, given a valid inequality $\sum_{v \in V} a_i x_i \leq a_0$ for the knapsack polytope P_K we may reduce the right-hand side a_0 by $\min\{a_u, a_v\}$.

To generalize this idea we define for $\bar{G} \subseteq G$ with $\bar{V} \subseteq V$, $\bar{E} \subseteq E_{\bar{V}}$ and $a \in \mathbb{R}_+^{|\bar{V}|}$ a function $\beta_{\bar{G}} : \{0,1\}^{|\bar{E}|} \to \mathbb{R} \cup \{\infty\}$ with

$$\beta_{\bar{G}}(y) = \inf\left\{ a(S), a(\bar{V} \setminus S) : S \subseteq \bar{V}, \max\left\{ a(S), a(\bar{V} \setminus S) \right\} \leq a_0, \, y = \chi^{\Delta_{\bar{G}}(S)} \right\}$$

and consider the convex envelope $\check{\beta}_{\bar{G}} : \mathbb{R}^{|\bar{E}|} \to \mathbb{R} \cup \{\infty\}$ of $\beta_{\bar{G}}(y)$, i.e.,

$$\check{\beta}_{\bar{G}}(x) = \sup\left\{ \check{\beta}(x) \, : \, \check{\beta} : \mathbb{R}^{|\bar{E}|} \to \mathbb{R}, \, \check{\beta} \text{ convex}, \, \check{\beta}(y) \leq \beta_{\bar{G}}(y) \text{ for } y \in \{0,1\}^{|\bar{E}|} \right\}. \quad (5.3)$$

Note that $\check{\beta}_{\bar{G}}$ is a piecewise linear function on its domain.

Definition 5.6 *Let $\sum_{v \in V} a_v x_v \leq a_0$ with $a_v \geq 0$ for all $v \in V$ be a valid inequality for the knapsack polytope P_K. Let V_0 be a non-empty subset of V and $r \in V_0$. Select subgraphs $(V_l, E_l) = G_l \subset G$ with pairwise disjoint sets V_l, $V_l \cap V_0 = \emptyset$ and $E_l \subseteq E(V_l)$ for $l = 1, \ldots, L$. Find for each l an affine minorant for the convex envelope $\check{\beta}_{G_l}$ such that*

$$c_0^l + \sum_{e \in E_l} c_e y_e \leq \check{\beta}_{G_l}(y) \quad (5.4)$$

holds for all y in P_B. Then for even $|H_v|$, $v \in V_0 \setminus \{r\}$, the inequality

$$\sum_{v \in V_0} a_v \left(1 - \sum_{e \in E_{rv} \setminus H_v} y_e - \sum_{e \in E_{rv} \cap H_v} (1 - y_e) \right) \leq a_0 - \sum_{l=1}^{L} (c_0^l + \sum_{e \in E_l} c_e y_e) \quad (5.5)$$

is called the **capacity reduced even bisection knapsack path inequality** *and for odd $|H_v|$, $v \in V_0 \setminus \{r\}$, the inequality*

$$\sum_{v \in V_0} a_v \left(1 - \sum_{e \in E_{rv} \setminus H_v} y_e - \sum_{e \in E_{rv} \cap H_v} (1 - y_e) \right) \leq a_r + a_0 - \sum_{l=1}^{L} (c_0^l + \sum_{e \in E_l} c_e y_e) \quad (5.6)$$

is called the **capacity reduced odd bisection knapsack path inequality**.

Below we give an example how the capacity reduction strengthening works.

Example 5.7 For the graph depicted in Figure 5.2 and the knapsack inequality

$$x_1 + 2x_2 + 3x_3 + x_4 + x_5 + 4x_6 \leq 10$$

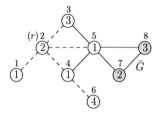

Figure 5.2: Capacity reduction based strengthening of an even bisection knapsack path inequality described in Example 5.7. The weights are within and the numbers of the nodes are above the circles.

defined on the tree (induced by the dashed edges) the even bisection knapsack path inequality

$$2 + (1 - y_{1,2}) + 3(1 - y_{2,3}) + (1 - y_{2,4}) + (1 - y_{2,5}) + 4(1 - (1 - y_{2,4}) - (1 - y_{4,6})) \le 10,$$

or equivalently

$$-y_{1,2} - 3y_{2,3} - 5y_{2,4} - y_{2,5} + 4y_{4,6} \le -2,$$

defines a zero-dimensional face of $P_{\mathcal{B}}$. Using the capacity of the graph $\bar{G}$ (induced by the gray nodes) we can reduce the right-hand side by $\min\{f_7, f_8\} = 2$ provided the edge $\{7, 8\}$ is in the cut. Thus the corresponding capacity reduced even bisection knapsack path inequality reads

$$-y_{1,2} - 3y_{2,3} - 5y_{2,4} - y_{2,5} + 4y_{4,6} \le -2 - 2y_{7,8}$$

and defines a face of $P_{\mathcal{B}}$ of dimension 3.

Proposition 5.8 (Armbruster 2007) *The capacity reduced bisection knapsack path inequalities (5.5) and (5.6) are valid for $P_{\mathcal{B}}$.*

Proof. Let $(S, V \setminus S)$ be a bisection in G with $r \in S$ and let $y = \chi^{\Delta(S)}$. Since

$$c_0^l + \sum_{e \in E_l} c_e y_e \le \check{\beta}_{G_l}(y) = \min \left\{ \sum_{v \in V_l \cap S} a_v, \ \sum_{v \in V_l \cap (V \setminus S)} a_v \right\},$$

we have

$$c_0^l + \sum_{e \in E_l} c_e y_e \le \sum_{V_l \cap S} a_v \quad \text{as well as} \quad c_0^l + \sum_{e \in E_l} c_e y_e \le \sum_{V_l \cap (V \setminus S)} a_v$$

for all $l = 1, \ldots, L$. Using the fact that $V_0, V_1, \ldots, V_L$ are pairwise disjoint sets we obtain

$$\sum_{v \in V_0} a_v \left(1 - \sum_{e \in P_{rv} \setminus H_v} y_e - \sum_{e \in H_v} (1 - y_e) \right) + \sum_{l=1}^{L} \left(c_0^l + \sum_{e \in E_l} c_e y_e \right)$$

$$\le \sum_{v \in V_0 \cap S} a_v + \sum_{l=1}^{L} \left(c_0^l + \sum_{v \in V_l \cap S} a_v \right) \le \sum_{v \in S} a_v \le a_0$$

for even $|H_v|$, $v \in V_0$, and

$$\sum_{v \in V_0} a_v \left(1 - \sum_{e \in P_{rv} \setminus H_v} y_e - \sum_{e \in H_v} (1 - y_e) \right) + \sum_{l=1}^{L} \left(c_0^l + \sum_{e \in E_l} c_e y_e \right)$$

$$\leq a_r + \sum_{v \in V_0 \setminus S} a_v + \sum_{l=1}^{L} \left(c_0^l + \sum_{v \in V_l \cap (V \setminus S)} a_v \right) \leq a_r + \sum_{v \in V \setminus S} a_v \leq a_r + a_0$$

for odd $|H_v|$, $v \in V_0$. See also Armbruster et al. [4, 7].   □

To find inequalities (5.4) needed in Proposition 5.8 one takes a closer look at the lower envelope defined in (5.3). In certain cases it is possible to establish a full description of $\breve{\beta}_{\bar{G}}$ by giving a complete description of the cluster weight polytope defined below.

Definition 5.9 *Let a graph $G = (V, E)$ and non-negative node weights $a_v \in \mathbb{R}$ for all $v \in V$ be given. For a set $S \subseteq V$ we define*

$$h(S) := \begin{pmatrix} a(S) \\ \chi^{\Delta(S)} \end{pmatrix} \in \mathbb{R}^{|E|+1}.$$

With respect to a given non-negative $a_0 \in \mathbb{R}$ we define

$$P_W(G) = \text{conv}\{ h(S) : S \subseteq V, a(S) \leq a_0, a(V \setminus S) \leq a_0 \}$$

and call this set the **cluster weight polytope**.

The purpose of studying $P_W(\bar{G})$ for some subgraph $\bar{G}$ of G is that its polyhedral description immediately yields the epigraph of $\breve{\beta}_{\bar{G}}$ via $\text{epi}(\breve{\beta}_{\bar{G}}) = P_W(\bar{G}) + \{\lambda(1, 0^T)^T : \lambda \geq 0\}$. This is the content of the next proposition.

Proposition 5.10 *Given a subgraph $\bar{G} = (\bar{V}, \bar{E})$ of G with node weights $a_v \geq 0$ for $v \in V$, an inequality of the form $y_0 + \sum_{e \in \bar{E}} \gamma_e y_e \geq \gamma_0$ is valid for $P_W(\bar{G})$ if and only if $\gamma_0 - \sum_{e \in \bar{E}} \gamma_e y_e$ is an affine minorant of $\breve{\beta}_{\bar{G}}$.*

Proof. The inequality $y_0 + \sum_{e \in \bar{E}} \gamma_e y_e \geq \gamma_0$ is valid for $P_W(\bar{G})$ if and only if $y_0 + \sum_{e \in \bar{E}} \gamma_e y_e \geq \gamma_0$ holds for

$$\begin{pmatrix} y_0 \\ y \end{pmatrix} \in \{ h_{\bar{G}}^S : S \subseteq \bar{V}, a(S) \leq a_0, a(\bar{V} \setminus S) \leq a_0 \}.$$

This is satisfied if and only if $y_0 \geq \gamma_0 - \sum_{e \in \bar{E}} \gamma_e y_e$ holds for

$$\left\{ \begin{pmatrix} \min\{a(S), a(\bar{V} \setminus S)\} \\ \chi^{\delta_{\bar{G}}(S)} \end{pmatrix} : S \subseteq \bar{V}, \max\{a(S), a(\bar{V} \setminus S)\} \leq a_0 \right\}.$$

Using the definition of $\beta_{\bar{G}}$ and the fact that inf $\emptyset = \infty$ we obtain that the latter statement is in turn equivalent to the validity of

$$\beta_{\bar{G}}(y) \geq \gamma_0 - \sum_{e \in \bar{E}} \gamma_e y_e$$

for $y \in \{0,1\}^{|\bar{E}|}$. Finally, due to the definition of $\check{\beta}_{\bar{G}}$

$$\check{\beta}_{\bar{G}}(y) \geq \gamma_0 - \sum_{e \in \bar{E}} \gamma_e y_e$$

holds for $y \in \mathbb{R}^{|\bar{E}|}$.   □

Hence, the "lower" facets of P_W are in one to one correspondence to the linear components of $\check{\beta}$. For a star $\bar{G} = (\bar{V}, \bar{E})$ we are able to exhibit facets of $P_W(\bar{G})$ which in certain problems enable us to strengthen bisection knapsack path inequalities of P_B to facet-defining inequalities of P_B.

Theorem 5.11 (Armbruster 2007) *Let $G = (V, E)$ be a star with center $r \in V$, $a \geq 0$, and $a(V) \leq a_0$. Then the following assertions hold.*

(a) The trivial inequalities

$$0 \leq y_{rv} \leq 1, \quad \forall \, v \in V \setminus \{r\}, \tag{5.7}$$

are facet-inducing for $P_W(G)$ except for one particular case: if there is exactly one $u \in V \setminus \{r\}$ with $a_u = a_r = \frac{1}{2}a(V)$ then $y_{ru} \leq 1$ does not induce a facet.

(b) Assume that $a(V \setminus \{r\}) > a_r$. Let $u \in V$ and $V_p \subset V$ such that $r, u \notin V_p$ and

$$a(V_p) \leq \tfrac{1}{2}a(V) < a(V_p) + a_u.$$

Let $V_n := V \setminus (V_p \cup \{r, u\})$. Then the inequalities

$$y_0 + \sum_{v \in V_p} a_v y_{rv} + (a(V) - 2a(V_p) - a_u)\, y_{ru} - \sum_{v \in V_n} a_v y_{rv} \;\leq\; a(V), \tag{5.8}$$

$$y_0 - \sum_{v \in V_p} a_v y_{rv} - (a(V) - 2a(V_p) - a_u)\, y_{ru} + \sum_{v \in V_n} a_v y_{rv} \;\geq\; 0 \tag{5.9}$$

are facet-inducing for $P_W(G)$.

(c) If $a(V \setminus \{r\}) \leq a_r$ then the inequalities

$$y_0 + \sum_{v \in V \setminus \{r\}} a_v y_{e_v} \;\leq\; a(V), \tag{5.10}$$

$$y_0 - \sum_{v \in V \setminus \{r\}} a_v y_{e_v} \;\geq\; 0 \tag{5.11}$$

are facet-inducing for $P_W(G)$.

For the proof we refer to Armbruster et al. [4, 7].

In the next theorem we show that the facets listed in Theorem 5.11 completely describe the cluster weight polytope P_W. Note that the assumption $a(V) \leq a_0$ guarantees that every $S \subseteq V$ contributes its point $h(S)$ to P_W. If we reduce a_0 below $a(V)$ then the facial structure of P_W becomes much more complicated, because then the whole complexity of the knapsack polytope P_K comes into play. So far a complete description of P_W with $a(V) > a_0$ seems out of reach, even if we assume $a_v = 1$ for all $v \in V$.

Theorem 5.12 (Armbruster et al. [4, 7]) *Let $G = (V, E)$ be a star with center $r \in V$, $a \geq 0$, and $a(V) \leq a_0$. Then the following assertions hold.*

(a) $P_W = \{y \in \mathbb{R}^{|E|+1} : y$ *fulfills* (5.7), (5.8) *and* (5.9)$\}$ *if* $a(V \setminus \{r\}) > a_r$.

(b) $P_W = \{y \in \mathbb{R}^{|E|+1} : y$ *fulfills* (5.7), (5.10) *and* (5.11)$\}$ *if* $a(V \setminus \{r\}) \leq a_r$.

The first proof of the above result was formulated by Armbruster [4]. Below we prove Theorem 5.12 in an alternative way. We refer also to Armbruster et al. [7]. First we show some intermediate results to finally give the proof of the theorem.

The general idea of our proof of Theorem 5.12 bases on the observation that all possible facets of P_W fall into one of the following three classes:

$$y_0 + \sum_{v \in V \setminus \{r\}} \gamma_v y_{rv} \leq \gamma_0, \tag{5.12}$$

$$\sum_{v \in V \setminus \{r\}} \gamma_v y_{rv} \leq \gamma_0, \tag{5.13}$$

$$-y_0 + \sum_{v \in V \setminus \{r\}} \gamma_v y_{rv} \leq \gamma_0. \tag{5.14}$$

Note that the facets of the form (5.12) bound the polytope P_W from the "upper side". Hence the points $h(S) \in P_W$, $S \subseteq V$, such that $h_0 = a(S) \geq a(V \setminus S)$ lie on the face induced by (5.12).

Lemma 5.13 *For an arbitrary facet of P_W of the form (5.12) we have*

$$\forall v \in V \setminus \{r\} : \quad -a_v \leq \gamma_v \leq a_v. \tag{5.15}$$

Proof. We assume first $\gamma_u > 0$ for some $u \in V$. Observe that (5.12) has a root satisfying $y_{ru} = 0$, otherwise all roots of (5.12) would satisfy the equation $y_{ru} = 1$, and thus our inequality could not induce a facet. Let

$$\begin{pmatrix} \hat{y}_0 \\ \hat{y} \end{pmatrix} = \begin{pmatrix} a(S) \\ \chi^{\Delta(S)} \end{pmatrix}$$

for an $S \subseteq V$ with $a(S) \geq a(V \setminus S)$ be a root of (5.12) such that $\hat{y}_{ru} = 0$. To bound γ_u we consider the cut $\Delta(S) \cup \{ru\}$ and distinguish the following cases concerning the location of node u.

(1) $u \in S$.

  (a) $a(S \setminus \{u\}) \geq a((V \setminus S) \cup \{u\})$.

  Consider $(\bar{y}_0, \bar{y}^T)^T = h(S \setminus \{u\}) \in P_{\mathcal{W}}$. Then

  $$\gamma_0 \geq \bar{y}_0 + \sum_{v \in V \setminus \{r\}} \gamma_v \bar{y}_{rv}.$$

  Since $(\hat{y}_0, \hat{y}^T)^T$ is a root of (5.12), we have

  $$\gamma_0 = \hat{y}_0 + \sum_{v \in V \setminus \{r\}} \gamma_v \hat{y}_{rv}.$$

  Thus

  $$\hat{y}_0 + \sum_{v \in V \setminus \{r\}} \gamma_v \hat{y}_{rv} \geq \bar{y}_0 + \sum_{v \in V \setminus \{r\}} \gamma_v \bar{y}_{rv},$$

  i.e., $\hat{y}_0 \geq \bar{y}_0 + \gamma_u$, and hence

  $$\gamma_u \leq \hat{y}_0 - \bar{y}_0 = a(S) - a(S \setminus \{u\}) = a_u.$$

  (b) $a(S \setminus \{u\}) < a((V \setminus S) \cup \{u\})$.

  This implies $a(S \setminus \{u\}) < \frac{1}{2}a(V)$. Consider $(\bar{y}_0, \bar{y}^T)^T = h((V \setminus S) \cup \{u\}) \in P_{\mathcal{W}}$. Using the reasoning from the previous case we obtain again

  $$\hat{y}_0 \geq \bar{y}_0 + \gamma_u.$$

  Therefore

  $$\gamma_u \leq \hat{y}_0 - \bar{y}_0 = a(S) - a((V \setminus S) \cup \{u\}) = a_u + 2a(S \setminus \{u\}) - a(V) < a_u,$$

  where the last inequality uses $a(S \setminus \{u\}) - \frac{1}{2}a(V) < 0$.

(2) $u \in V \setminus S$.

We show that this case does not occur. By the assumption $a(S) \geq a(V \setminus S)$, thus $a(S \cup \{u\}) \geq a(V \setminus (S \cup \{u\}))$. We take $(\bar{y}_0, \bar{y}^T)^T = h(S \cup \{u\}) \in P_{\mathcal{W}}$. As in case 1(a) we obtain

$$\hat{y}_0 \geq \bar{y}_0 + \gamma_u.$$

Therefore

$$\gamma_u \leq \hat{y}_0 - \bar{y}_0 = a(S) - a(S \cup \{u\}) = -a_u.$$

This contradicts our assumption $\gamma_u > 0$, thus the case $u \in V \setminus S$ is not possible.

To show that $-a_u \leq \gamma_u$ in case $\gamma_u < 0$ one can use an analogous argumentation as above. This time one considers a root of (5.12) satisfying $y_{ru} = 1$ and the cut $\Delta(S) \setminus \{ru\}$. $\quad\square$

Lemma 5.14 *For an arbitrary facet of P_W of the form (5.12) we have*

$$\gamma_0 = a(V), \tag{5.16}$$

$$\sum_{v \in V \setminus \{r\}} \gamma_v \leq a_r. \tag{5.17}$$

Proof. Since (5.12) has to be satisfied by $h(V) \in P_W$, we immediately obtain

$$a(V) \leq \gamma_0.$$

If $a_r \geq a(V \setminus \{r\})$ then (5.12) has to be satisfied by $h(\{r\}) \in P_W$ and we get

$$\sum_{v \in V \setminus \{r\}} \gamma_v \leq \gamma_0 - a_r \leq \gamma_0 - a(V \setminus \{r\}).$$

If $a_r < a(V \setminus \{r\})$ then (5.12) has to be satisfied by $h(V \setminus \{r\}) \in P_W$. Thus also in this case we obtain

$$\sum_{v \in V \setminus \{r\}} \gamma_v \leq \gamma_0 - a(V \setminus \{r\}). \tag{5.18}$$

It is enough to show that (5.16) holds, since (5.16) together with the above inequality implies (5.17). Let $(\bar{y}_0, \bar{y}^T)^T = h(S) \in P_W$, $S \subseteq V$ such that $a(S) \geq a(V \setminus S)$. If $r \in S$ then

$$\bar{y}_0 + \sum_{v \in V \setminus \{r\}} \gamma_v \bar{y}_{rv} = a(S) + \sum_{v \in V \setminus S} \gamma_v \leq a(S) + \sum_{v \in V \setminus S} a_v = a(V),$$

where the inequality in the middle follows from (5.15). Hence, if $\gamma_0 > a(V)$ none of these points lies on the face induced by (5.12). We show that all points lying on the face induced by this inequality, i.e., points $h(S)$ with $r \in V \setminus S$, also satisfy another valid inequality for P_W with equality. By (5.18) we have

$$\sum_{v \in V \setminus \{r\}} (a_v + \gamma_v) \leq \gamma_0. \tag{5.19}$$

Since $a_v + \gamma_v \geq 0$ by Lemma 5.13 and $y_{rv} \in [0, 1]$ for all $(y_0, y^T)^T \in P_W$, we conclude that

$$\sum_{v \in V \setminus \{r\}} (a_v + \gamma_v) y_{rv} \leq \gamma_0 \tag{5.20}$$

is a valid inequality for P_W. Note that (5.20) is not a scalar multiple of (5.12). Now, if

$$\bar{y}_0 + \sum_{v \in V \setminus \{r\}} \gamma_v \bar{y}_{rv} = \gamma_0$$

holds then

$$\bar{y}_0 = \sum_{v \in V \setminus \{r\}} a_v \bar{y}_{rv}$$

and we obtain

$$\gamma_0 = \sum_{v \in V \setminus \{r\}} (a_v + \gamma_v) \bar{y}_{rv}.$$

Thus all points of P_W which lie on the face induced by (5.12) also lie on the face induced by (5.20). Therefore (5.12) cannot be a facet of P_W if $\gamma_0 \neq a(V)$.   □

Proposition 5.15 *For a star $G = (V, E)$ with root $r \in V$, $a \geq 0$, and $a(V) \leq a_0$ all facets of the form (5.12) for P_W are defined by (5.8) if $a(V \setminus \{r\}) > a_r$ and by (5.10) if $a(V \setminus \{r\}) \leq a_r$.*

Proof. We only need to determine the best γ_0 and γ subject to the constraints (5.15) - (5.17) so that

$$\gamma_0 - \sum_{v \in V \setminus \{r\}} \gamma_v y_{rv} - y_0$$

is as small as possible for any given $y \in [0, 1]^{|E|}$. In other words, we need to solve the following problem:

$$\min \quad a(V) - \sum_{v \in V \setminus \{r\}} y_{rv} \gamma_v,$$

$$\text{s.t.} \quad \sum_{v \in V \setminus \{r\}} \gamma_v \leq a_r, \tag{5.21}$$

$$-a_v \leq \gamma_v \leq a_v, \quad \forall v \in V \setminus \{r\}.$$

Substituting $\tilde{\gamma}_v = \gamma_v + a_v$ we transform the above problem to

$$\max \quad \sum_{v \in V \setminus \{r\}} y_{rv} \tilde{\gamma}_v - \sum_{v \in V \setminus \{r\}} y_{rv} a_v,$$

$$\text{s.t.} \quad \sum_{v \in V \setminus \{r\}} \tilde{\gamma}_v \leq a(V), \tag{5.22}$$

$$0 \leq \tilde{\gamma}_v \leq 2a_v, \quad \forall v \in V \setminus \{r\},$$

which is just a continuous bounded knapsack problem (see Remark A.21) with variables $\tilde{\gamma}_v$, $v \in V \setminus \{r\}$. Each item $v \in V \setminus \{r\}$ has the profit y_{rv}, the weight equal to 1, and the availability at most $2a_v$. The knapsack capacity is $a(V)$. One can find an optimal solution by sorting the items with respect to non-increasing profit-to-weight ratios y_{rv}, $v \in V \setminus \{r\}$. W.l.o.g., we can pack the knapsack in the order $1, 2, \ldots, |V| - 1$. If $a(V \setminus \{r\}) > a_r$, i.e., $a(V) > 2a_r$ then

$$\sum_{v \in V \setminus \{r\}} 2a_v = 2a(V) - 2a_r > a(V).$$

and thus not all items fit into the knapsack with their full availability $2a_v$, $v \in V \setminus \{r\}$. Hence there exists some item u such that

$$2(a_1 + \ldots + a_{u-1}) \leq a(V) \quad \text{and} \quad 2(a_1 + \ldots + a_u) > a(V).$$

Thus we obtain

$$
\begin{aligned}
\tilde{\gamma}_v &= 2a_v, \quad v = 1, \ldots, u - 1, \\
\tilde{\gamma}_u &= a(V) - 2(a_1 + \ldots + a_{u-1}), \\
\tilde{\gamma}_v &= 0, \quad v = u + 1, \ldots, |V| - 1.
\end{aligned}
$$

The solution to the problem (5.21) reads

$$
\begin{aligned}
\gamma_v &= a_v, \quad v = 1, \ldots, u - 1, \\
\gamma_u &= a(V) - 2(a_1 + \ldots + a_{u-1}) - a_u, \\
\gamma_v &= -a_v \quad v = u + 1, \ldots, |V| - 1.
\end{aligned}
$$

Now, taking $V_p = \{1, \ldots, u - 1\}$ we get inequality (5.8).

In the case of $a(V \setminus \{r\}) \le a_r$ all items can be packed with their full availability into the knapsack. Thus $\gamma_v = a_v$ for all $v \in V \setminus \{r\}$ and we obtain inequality (5.10). $\qquad\square$

Proposition 5.16 P_W *is symmetric with respect to the hyperplane*

$$
\{\, y \in \mathbb{R}^{|E|} : 2y_0 = a(V) \,\}.
$$

Proof. Consider the pairs of points $(h(S), h(V \setminus S))$ from P_W. Since $\chi^{\Delta(S)} = \chi^{\Delta(V \setminus S)}$, we have

$$
\begin{pmatrix} \frac{1}{2} a(V) \\ \chi^{\Delta(S)} \end{pmatrix} - h(S) = h(V \setminus S) - \begin{pmatrix} \frac{1}{2} a(V) \\ \chi^{\Delta(S)} \end{pmatrix}.
$$

$\qquad\square$

Proposition 5.17 *For a star $G = (V, E)$ with root $r \in V$, $a \ge 0$, and $a(V) \le a_0$ all facets of the form (5.14) for P_W are defined by (5.9) if $a(V \setminus \{r\}) > a_r$ and (5.11) if $a(V \setminus \{r\}) \le a_r$.*

Proof. Using the symmetry of P_W with respect to the hyperplane $\{\, y \in \mathbb{R}^{|E|} : 2y_0 = a(V) \,\}$ we obtain inequalities (5.14) from (5.12) with the same values for γ_v and γ_0. Similarly we get (5.9) from (5.8) and (5.11) from (5.10) and apply Proposition 5.15. $\quad\square$

Proposition 5.18 *For a star $G = (V, E)$ with root $r \in V$, $a \ne 0$ and $a(V) \le a_0$ all facets of the form (5.13) for P_W are defined by (5.7).*

Proof. Due to Proposition A.5 a facet of a polytope is also a facet of the projection of this polytope if one projects out the variable with the zero-coefficient in the inequality defining this facet. Since the variable y_0 has the coefficient equal to zero in the inequalities of the form (5.13), we look at the projection P' of P_W onto the space $\mathbb{R}^{|E|}$, i.e.,

$$P' := \{ y \in \mathbb{R}^{|E|} \: : \: \begin{pmatrix} y_0 \\ y \end{pmatrix} \in P_W \}.$$

We show that P' has only facets of the form $0 \leq y_{rv} \leq 1$, $v \in V \setminus \{r\}$. Due to the assumption $a(V) \leq a_0$ we have

$$P_W = \text{conv} \left\{ \begin{pmatrix} a(S) \\ \chi^{\Delta(S)} \end{pmatrix} \in \mathbb{R} \times \{0,1\}^{|E|} \: : \: S \subseteq V \right\}.$$

Thus

$$P' = \text{conv}\{\chi^{\Delta(S)} \in \{0,1\}^{|E|} \: : \: S \subseteq V \} = \text{conv} \{ y \in \{0,1\}^{|E|} \}.$$

Therefore P' is exactly the $|E|$-dimensional hypercube which is completely described by inequalities $0 \leq y_i \leq 1$, $i = 1, \ldots, |E|$. Hence all facets of P_W with the zero-coefficient at y_0 can only have the form (5.7) and the claim of the proposition follows. $\square$

Finally, we can turn to the proof of Theorem 5.12.

Proof of Theorem 5.12. If $a(V \setminus \{r\}) > a_r$ then

$$P_W \subseteq \{y \in \mathbb{R}^{|E|+1} : y \text{ fulfills } (5.7), (5.8), \text{ and } (5.9)\}$$

due to Theorem 5.11 (a) and (b).

$$P_W \supseteq \{y \in \mathbb{R}^{|E|+1} : y \text{ fulfills } (5.7), (5.8), \text{ and } (5.9)\}$$

follows from Propositions 5.15, 5.17, and 5.18.

If $a(V \setminus \{r\}) \leq a_r$ then

$$P_W \subseteq \{y \in \mathbb{R}^{|E|+1} : y \text{ fulfills } (5.7), (5.10), \text{ and } (5.11)\}$$

due to Theorem 5.11 (a), (c) and

$$P_W \supseteq \{y \in \mathbb{R}^{|E|+1} : y \text{ fulfills } (5.7), (5.10), \text{ and } (5.11)\}$$

follows again from Propositions 5.15, 5.17, and 5.18. $\square$

5.3   Separation

In our computational tests we consider the wider class of bisection knapsack path inequalities namely the bisection knapsack walk inequalities. The separation algorithm for the bisection knapsack walk is designed and described in details by Armbruster in [4]. Below we outline the leading ideas.

The number of all valid even and odd bisection knapsack walk inequalities is exponential in the dimension of the polytope P_B. This is due to the exponential number of valid inequalities for the knapsack polytope P_K, as well as due to the exponential number of possible paths joining any two nodes in the graph G. Another obstacle in separating the bisection knapsack walk inequalities, as in case of knapsack tree inequalities, is to find the underlying valid inequality[1] for the knapsack polytope P_K. Nevertheless, having a valid knapsack inequality for P_K we can identify in polynomial time a bisection knapsack walk inequality violating a given solution of a relaxation if one exists. This is possible due to the strong connection between bisection knapsack walk and odd cycle inequalities (3.16). Consider the term ω_{rv}, $r, v \in V$, corresponding to a path $\Pi_{rv} = (V_{rv}, E_{rv})$ joining nodes r and v, and assume that there exists an edge rv in G joining these two nodes. Since $(V_{rv}, E_{rv} \cup \{rv\})$ is a cycle, we apply the odd cycle inequality (3.17)

$$\sum_{e \in E_{rv} \setminus H_v} y_e + \sum_{e \in H_v \cup \{rv\}} (1 - y_e) \geq 1$$

for even $|H_v|$ and

$$\sum_{e \in (E_{rv} \cup \{rv\}) \setminus H_v} y_e + \sum_{e \in H_v} (1 - y_e) \geq 0$$

for odd $|H_v|$. Hence we obtain[2]

$$\omega_{rv} = 1 - \sum_{e \in E_{rv} \setminus H_v} y_e - \sum_{e \in H_v} (1 - y_e) \leq 1 - y_{rv}. \tag{5.23}$$

The relationship between the term ω_{rv} and the left-hand side of the odd cycle inequality exposed above allows us to apply the algorithm of Barahona and Mahjoub [11], described in Section 3.4, in the separation procedure for bisection knapsack walk inequalities, outlined in Algorithm 5.1. Let $\bar{y}$ be a solution of a relaxation to the minimum bisection problem and let $\sum_{i \in V} a_i x_i \leq a_0$ with $a_i > 0$, $i \in V'$, be a valid inequality for P_K. We aim to find a bisection knapsack walk inequality violated by $\bar{y}$. Hence for some $r \in V'$ and each $v \in V'$ we need to establish the paths Π_{rv} and sets $H_v \subseteq E_{rv}$ such that

$$\sum_{v \in V'} a_v \left(1 - \sum_{e \in E_{rv} \setminus H_v} \bar{y}_e - \sum_{e \in H_v} (1 - \bar{y}_e) \right) < a_0$$

for even $|H_v|$ (Step 11) and

$$\sum_{v \in V'} a_v \left(1 - \sum_{e \in E_{rv} \setminus H_v} \bar{y}_e - \sum_{e \in H_v} (1 - \bar{y}_e) \right) < a_r + a_0$$

for odd $|H_v|$ (Step 21). For convenience we set $\tilde{E}_{rv} := E_{rv} \setminus H_v$ in Step 17.

---

[1]So far the knapsack inequality defining the bisection problem $\sum_{v \in V} f_v x_v \leq u_r$ as well as cover knapsack inequalities (A.23) are applied, see also Section 4.3.

[2]Armbruster [4] uses (5.23) to prove the validity of bisection knapsack walk inequalities for P_B.

Algorithm 5.1 Separation Algorithm for Bisection Knapsack Walk Inequalities

---

Input Current solution $\bar{y}$ of a relaxation to the MGBP.

1:   $i = 0$ counter for supports violating the odd cycle inequality;

2:   create auxiliary graph $\bar{G} = (\bar{V}, \bar{E})$ from $G = (V, E)$:

3:      for each $v \in V$ add node $\bar{v}$;

4:      for each $vu \in E$ add edges $\bar{v}u$, $v\bar{u}$, and $\bar{v}\bar{u}$;

5:      set edge weights $\varphi_{vu} = \varphi_{\bar{v}\bar{u}} = \bar{y}_{vu}$, $\varphi_{\bar{v}u} = \varphi_{v\bar{u}} = 1 - \bar{y}_{vu}$;

6: **for each** $r \in V'$ **do**

7:     **for each** $v \in V'$ **do**

8:        compute shortest paths $\Pi_{rv} = (\bar{V}_{rv}, \bar{E}_{rv})$ joining r, v and $\Pi_{r\bar{v}} = (\bar{V}_{r\bar{v}}, \bar{E}_{r\bar{v}})$ joining r, $\bar{v}$ in $\bar{G}$ with respect to weights φ;

9:     **end for**

10:     find a knapsack inequality $a^T x \leq a_0$, $a_i > 0$, $i \in V'$;

11:     **if** $\sum_{v \in V'} a_v \varphi(\bar{E}_{rv}) > a_0$ **then**

12:        $i \to i + 1$,   mark i as **even**;

13:        **for each** $v \in V'$ and $e \in \bar{E}_{rv}$ **do**

14:           **if** $e = t\bar{u}$ or $e = \bar{t}u$, $(t \in V, \bar{u} \in \bar{V} \setminus V)$ **then**

15:             $tu \to H_v^i$;

16:           **else**

17:             $tu \to \tilde{E}_{rv}^i$;

18:           **end if**

19:        **end for**

20:     **end if**

21:     **if** $\sum_{v \in V'} a_v \varphi(\bar{E}_{r\bar{v}}) > a_r + a_0$ **then**

22:        $i \to i + 1$,   mark i as **odd**;

23:        **for each** $v \in V'$ and $e \in \bar{E}_{r\bar{v}}$ **do**

24:           repeat steps 14 - 18;

25:        **end for**

26:     **end if**

27: **end for**

Output If $i > 0$, list of supports $(\tilde{E}_{rv}^1, H_v^1)_{v \in V'}, \ldots, (\tilde{E}_{rv}^i, H_v^i)_{v \in V'}$ yielding violated even and odd (resp.) bisection knapsack walk inequalities.

Obviously, the bigger the gap between the value of the left- and the right-hand side the better the selection of the support. Hence we look for the smallest possible values of ω_{rv}, $v \in V'$. For each $v \in V'$ we identify sets $\tilde{E}_{rv} := E_{rv} \setminus H_v$ and H_v using the auxiliary graph explained in Step 2 - 5 in Algorithm 5.1. If $|H_v|$ should be even, we look for a shortest path from r to v, i.e., we cross edges joining original and auxiliary nodes an even number of times. Similarly, should $|H_v|$ be odd, we compute a shortest path joining r and $\bar{v}$, the copy of v, thus we traverse an odd number of times between original and auxiliary nodes (Step 8). Since we allow walks, we do not forbid passing by nodes already visited. To obtain all violated inequalities we consider each $r \in V'$. Thus the algorithm is of complexity at most $\mathcal{O}(|V|^3)$. Note that we find the terms ω_{rv}, $v \in V'$, independently from the knapsack inequality $\sum_{i \in V} a_i x_i \leq a_0$. First we establish the shortest walks and then apply known separation algorithms for the knapsack polytope to find the coefficients a_i, $i \in V'$, (Step 10).

The algorithmic issues of the capacity reduction based strengthening (5.5) and (5.6) of bisection knapsack walk inequalities is too complex to outline it briefly. So far, the separation algorithm has been implemented only for graphs with all node weights equal to 1. We refer the reader to the original exposure in [4].

Chapter 6

Primal Heuristics

Primal heuristics are an important part of the bounding procedure in a branch-and-cut algorithm. They contribute to the reduction of the size of the branch-and-bound tree. In our minimization problem they deliver upper bounds on the objective value. If for a subproblem the lower bound is greater than some given upper bound then the feasible region of this subproblem cannot contain a better solution than that one corresponding to this upper bound. Hence we may delete this subproblem and its successors from the list of open problems. Obviously, the better the upper bound the more limited the search space and thus the shorter the solution time. Therefore we develop primal methods for solving the MGBP.

After a brief summary about known heuristics for various graph partitioning problems we describe three methods which we apply to find primal solutions for the MGBP. The first algorithm bases on an evolutionary approach. It is the only method, which we developed, that incorporates the solution of LP-relaxation to the MGBP to find upper bounds. The second heuristic applies a greedy randomized search procedure to construct a bisection and then improves the objective function value by swapping the elements in the clusters. We also use this method as an improving heuristic taking as the input the current best primal solution found during the branch-and-cut procedure. The third algorithm we adopted from the software packet METIS for partitioning graphs and adjusted for the special structure of the MGBP. This method follows purely the primal approach, i.e., it uses only the specification of the problem.

6.1 Known Heuristics for Graph Partitioning Problems

Graph partitioning problems are generally NP-hard [36]. Because of this and the large amount of applications many heuristics were developed. Among them there are algorithms which are devoted exclusively to partitioning problems sometimes narrowed to very special cases, as well as procedures which adapt known metaheuristics applicable to a wide class of optimization problems or even combinations of them called hybrid heuristics. However, most applications lead back to the k-partition problem concerning the partition the vertices of the graph into k roughly equal sized subsets such that the number of edges connecting vertices in different subsets is minimized.

Since its publication more than three decades ago the **Kernighan-Lin** heuristic [58] turned out to be the dominating procedure for finding good solution for the 2-partition problem. The heuristic starts with an arbitrary initial partition and improves the objective value by exchanging the pairs of nodes in different clusters. The nodes with the highest value of a greedy gain function are selected for a swap. Over the years many improvements and variations of the Kernighan-Lin heuristic have been developed. For example, Fiduccia and Mattheyes [33] improve the running time of the algorithm by using a heap data structure to order the gains. Elrod, Feo, and Laguna [28] introduce a randomized construction of the initial partition as well as additional swapping criteria. We adapt this technique for solving the MGBP in Section 6.3 below.

The Kernighan-Lin algorithm is also applied in heuristics based on the **multilevel** paradigm, see e.g. Karypis and Kumar [56] and Cross and Walshaw [22]. Such methods consist of three phases: graph coarsening, initial partitioning and multilevel refinement. In the coarsening phase, a series of graphs is constructed by collapsing together selected vertices of the input graph. This newly constructed graph acts as the input graph for another round of graph coarsening and so on, until a sufficiently small graph is obtained. The final graph is partitioned. Then partition refinement is performed on each coarsening level using the Kernighan-Lin algorithm till the original graph is reached. The algorithms implemented within the software METIS, developed by Karypis and Kumar [56], are based on the multilevel paradigm. In Section 6.4 we explain how this package is used to solve the MGBP.

Berry and Goldberg [13] develop a method called **path optimization**. The procedure is a sort of the hill-climbing local optimization. Given an initial partitioning it performs a variation of the simple neighborhood search. However, instead of selecting and moving a small constant number of vertices, as most local search methods do, the path optimization develops variable-length sequences of adjacent vertices which are moved to the respective opposite partition.

Another method for solving the 2-partition problem bases on a discrete quadratic programming formulation. Here, the key role plays the Laplacian L of the underlying graph. Assuming that the graph is connected the magnitude of the second largest eigenvalue of L gives a measure of the connectivity of the graph. The components of the eigenvector corresponding to this eigenvalue (known as Fiedler vector) reflect the connectivity between the vertices of the graph. They are sorted by this value and the ordered list is split into two parts producing a 2-partition. Although such **spectral techniques** are known to deliver good partitions they are computationally very expensive due to the calculation of the Fiedler vector.

There is a series of publications on the solution of graph partitioning problems with metaheuristics. Bui, Heighan, Jones, and Leighton [15], as well as Johnson, Aragon, McGeoch, and Schevon [52] apply the **simulated annealing** approach. It attempts to find a good solution by starting with an arbitrary initial partition and improving it through neighborhood search based on pairwise exchanges. The basic simulated annealing algorithm is a local optimization heuristic that allows even non-improving moves with a certain probability which decreases with a *temperature* according to some *cooling schedule*. The quality of the final solution, like the physical analogy, is better when the cooling occurs gradually rather than suddenly. When solving graph partitioning problems the temperature controls the number of node exchanges in each cooling step.

Rolland, Pirkul, and Glover [79] develop **tabu search** for the 2-partition problem. Starting with a randomly selected bisection they iteratively search for a better solution. In each iteration a node is selected and moved to the other set of the partition. After the move the node is kept in a so called tabu list for a certain number of iterations. However, a move of a node from the tabu list is allowed if it leads to a bisection with an improved cut weight. From all feasible moves the algorithms selects those moves that yield the greatest reduction of the cut weight. Dell'Amico and Maffili [26] apply tabu search for the 2-partition problem.

A class of metaheuristics widely used for solving graph partitioning problems is based on the **evolutionary search**. In the next section we treat these algorithms more profoundly and describe our implementation of such a heuristic for the MGBP.

An example of a **hybrid heuristic** is presented by Banos, Gil, Ortega, and Montoya [9]. They hybridize tabu search and simulated annealing. The heuristic applies at each iteration a local search. The outer structure replicates simulated annealing, admissible moves are applied with the probability steered by the cooling temperature. When a move improving the cost function is accepted, the reverse move is forbidden for several iterations to avoid cycling, as in tabu search. Battiti and Bertosi [12] describe an implementation of tabu search with a randomized number of prohibited moves and the initial solution is established by a greedy method similar to Elrod et al. [28].

We refer to Fjällström [34] and Schloegel, Karypis, and Kumar [80] for more detailed surveys on heuristic algorithms for graph partitioning problems.

6.2 Genetic Algorithm

Genetic algorithms solve optimization problems in an analogous manner to the evolution process of nature, see for instance Hoose and Stützle [49] and Michalewicz [71]. A solution of a given problem is coded into a string of **genes** termed **individual**. New solutions are generated by operations called **crossover** and **mutation**. In a crossover two **parents** are drawn from the current population and parts of their genes are exchanged resulting in **child** solutions. A mutation is an adequate transformation of a single individual. Individuals delivering the best objective value are selected for creating the next generation. All operations, creating the initial population of individuals, mutations, and crossover, are randomized.

In the applications for graph partitioning problems the individual is a 0-1 or integer (depending on the number of partitions) vector of the length $|V|$. The vertices corresponding to components of this vector with equal values are assigned to one cluster. The weight of the cut corresponding to the partition represented by an individual describes the individual's fitness, additionally penalized if the partition is not feasible.

Genetic algorithms are known to find good solutions to graph partitioning problems. The approaches vary on implementations of the mutation and crossover operations. Frequently local improvement heuristics are applied. Maini, Mehrotra, Mohan, and Ranka [68] present a genetic algorithm for k-partitioning. There, the crossover operation is improved by utilizing information available from the history of genetic search. Bui and Moon [16] also solve k-partitioning problems with an evolutionary method. A local improvement, based on a Kernighan-Lin approach, is performed after a new individual has been created by a mutation or a crossover. Kohmoto, Katayaman, and Narihisa [60] apply a genetic algorithm for solving the 2-partition problem. The fitness of the individuals in the initial population is improved by a local search based on the exchange of nodes from different clusters. There are also successful methods achieved by combining genetic algorithms with other classical metaheuristics. For instance, Soper, Walshaw, and Cross [83] hybridize evolutionary search with a multilevel heuristic.

The wide and effective application of evolutionary methods for graph partitioning problems motivated us to incorporate a primal heuristic based on a genetic algorithm in our branch-and-cut framework for solving the MGBP. We present a sort of a primal-dual hybrid method. Namely, besides the primal information, i.e., the edge costs, we exploit the solution of the LP-relaxation. Puchinger and Raidl give in [75] a classification of methods that combine exact and heuristic procedures. Within this scheme our work can be categorized as a metaheuristic for obtaining incumbent solutions and bounds. The ideas presented below can be also found in [5, 6].

In our heuristic, outlined in Algorithm 6.1, we code an individual in two equivalent ways. One form is the **node representation**, i.e., as a vector $v \in \{0,1\}^n$, $n = |V|$. If $v_i = v_j$, nodes i and j are in one cluster. The other form is the **edge representation** as a vector $y \in \{0,1\}^{|E|}$, where

$$y_{ij} = \begin{cases} 0, & v_i = v_j, \\ 1, & v_i = 1 - v_j. \end{cases}$$

To obtain more variety in the selection of individuals we allow infeasible solutions in the sense that the bounds l_τ and u_τ on the total weight node of clusters do not need to be fulfilled.

Algorithm 6.1 Genetic Heuristic

Input Current solution $\bar{y}$ of LP-relaxation, list of best primal solutions L, graph G, parameters: number of individuals in population p, factor of population growth k, total number of generations g, number of generations without fitness improvement f.

1: $n_g := 0$ (counter for all loops);

2: $n_f := 0$ (counter for loops without fitness improvement);

3: based on $\bar{y}$ and solutions from L create initial population with p individuals;

4: evaluate fitness of individuals;

5: **repeat**

6: perform crossovers and mutations till population grows to kp individuals;

7: evaluate fitness of new individuals;

8: **if** the best fitness is not improved **then** $n_f := n_f + 1$;

9: form the population of p best individuals;

10: $n_g := n_g + 1$;

11: **until** $n_g \geq g$ **or** $n_f \geq f$

Output Individual with the best fitness value.

The main idea of our genetic algorithm is to construct solutions to the bisection problem using the vector $\bar{y} \in [0, 1]^{|E|}$ which is a fractional solution to LP-relaxation of the MGBP. As an additional criterion we use the edge cost vector w. Thus we combine the components of $\bar{y}$ and w by considering $y^h = f(\bar{y}, w)$. The components y^h_{ij}, $ij \in E$ are supposed to provide the information if i and j should be assigned to the same cluster. The hybridization forms of y^h vary on subroutines of the algorithm (creating the starting population and mutations) and will be explained sequentially in the next sections.

Initial population

The initial population is determined in the following way. Suppose we are given a set $F \subseteq E$. How F is constructed we explain below. We compute a spanning forest on F such that for each of its components $(V^1, E^1), \ldots, (V^k, E^k) \subset (V_F, E_F)$, $k \geq 2$, holds:

$$\sum_{i \in V^j} f_i \leq u_\tau, \quad 1 \leq j \leq k.$$

The two "heaviest" sets, say V^s, V^t for some $1 \leq s, t \leq k$, with respect to the weighted node sum form the initial clusters. Using a bin packing heuristic (see Section A.5) we complete the clusters with the node sets V^j, $1 \leq j \leq k$, $j \neq s, t$. Thus we obtain an individual $\tilde{v} \in \{0, 1\}^n$ with $\tilde{v}_i = 0$ for all $i \in V^s$ and $\tilde{v}_i = 1$ for all $i \in V^t$. The above procedure can be seen as a modified Kruskal's algorithm (see Algorithm A.3) based on the heuristic *Edge* presented by Ferreira et al. in [32].

We construct F due to the information delivered by the fractional LP-solution $\overline{y}$ and the edge cost vector w. Concerning $\overline{y}$ the set F should contain edges with LP-values close to zero. In regard to w, F should possibly contain edges with high costs. We applied three methods of combining components of $\overline{y}$ and w. We consider $F := F_\varepsilon \cup F_\rho$, where

$$F_\varepsilon = \{e \in E : \overline{y}_e < \varepsilon\},$$

and F_ρ takes one of the following forms:

(F1) $\{e \in E : \frac{1}{1+w_e^2} \leq X_\rho\}$,

(F2) $\{e \in E : \overline{y}_e + \frac{1}{1+w_e^2} \leq X_\rho\}$,

(F3) $\{e \in E : Z\overline{y}_e + (1-Z)\frac{1}{1+w_e^2} \leq X_\rho\}$ for a random number $Z \in [0,1]$,

where X_ρ is either constantly equal to ρ or a random number in $[0, \rho]$. In (F1) an edge enters F if it has high cost and also a high LP-value with respect to ε. In (F2) we allow the extension of F by edges with high cost if their LP-value is not too high. In (F3) we prioritize randomly the approach (F1) and (F2) using a parameterization of the corresponding components of $\overline{y}$ and w. As a comparison to the above hybridization of the set F we consider also a random approach:

($\mathcal{R}$) $F_\rho = \{e \in E : \overline{y}_e < X_\rho$ and $\frac{\overline{y}_e}{X_\rho} < Z\}$ for a random $Z \in [0, 1]$.

Note that the bounds ε and ρ control the number of edges entering F. If $F_\rho = \emptyset$ then the starting population contains p duplicates of $\tilde{v}$. Otherwise, we select a new random value X_ρ in each iteration creating a new individual. In this way we obtain an initial population with different individuals.

In the latest version of the algorithm we merge into the initial population several individuals corresponding to the best upper bounds previously found by other heuristics or relaxations. Thus we use the genetic algorithm as an improving heuristic. As usual this happens at some random decision.

Mutations

To create a new generation we apply four mutation types. The parameter m_r called **mutation rate** is introduced to control the percentage of exchanged components of vector v or y, respectively, in one mutation round. We apply also hybridization methods to some of our mutations procedures. Concerning the implementation issues, we were able to apply the edge costs to Mutation 3 and 4 (see below) without loss on the efficiency of our algorithm. We implemented the following transformations.

Mutation 1. Let $v \in \{0,1\}^{|V|}$ be an individual selected for the mutation. Let $V_M \subset V$ be a randomly selected subset of v components such that $|V_M| < m_r$. For all $i \in V_M$ we set $v_i := 1 - v_i$. Note that this general procedure does not make use of the specifics of the underlying bisection problem.

Mutation 2. This procedure works in a similar way as Mutation 1 applied to the edge representation of the individual. Let $E_M \subset E$ be a random selection of edges such that $|E_M| < m_r$. For each $ij \in E_M$ such that $y_{ij} := 1$ we set $y_{ij} := 0$. Then we accordingly update the values of v_i and v_j. It holds $v_i \neq v_j$. Under a random decision we set either $v_i := v_j$ or $v_j := v_i$.

Mutation 3. Here we use the same idea as for the creation of the initial population. Let $E_M \subset E$ be a random selection of edges such that $|E_M| < m_r$ and let $X \in [0,1]$ be a random number. We consider the value $p_{ij} := 1 - y_{ij}^h$ as the probability that nodes i and j are in the same cluster, where

$$y_{ij}^h := Z\overline{y}_{ij} + \frac{1 - Z}{1 + w_{ij}},$$

for $ij \in E_M$ and a random $Z \in [0,1]$. For each $ij \in E_M$ if $p_{ij} > 1 - X$ we set either $v_i := v_j$ or $v_j := v_i$ at random. If $p_{ij} < X$ then we set $v_i := 1 - v_j$ or $v_j := 1 - v_i$, again randomly.

Mutation 4. This is a kind of neighborhood search. Let $v \in \{0,1\}^{|V|}$ be an individual selected for the mutation. Let $V_M \subset V$ be a randomly selected subset of v components such that $|V_M| < m_r$. For $i \in V_M$ we count the number of nodes adjacent to i in each cluster, i.e., we determine the numbers

$$c_0 = |\{j : ij \in E, v_j = 0\}| \quad \text{and} \quad c_1 = |\{j : ij \in E, v_j = 1\}|.$$

If $c_0 > c_1$ we set $v_i := 0$, and $v_i := 1$ otherwise. Furthermore, we sum up the edge costs of joining node $i \in V_M$ with nodes in each cluster separately, i.e., we calculate

$$w_0 = \sum_{ij \in E : v_j = 0} w_{ij} \quad \text{and} \quad w_1 = \sum_{ij \in E : v_j = 1} w_{ij}.$$

We decide randomly, either v_i takes its value depending on $\max\{c_0, c_1\}$, as in the original procedure, or depending on $\min\{w_0, w_1\}$, i.e., if $w_0 < w_1$ we set $v_i := 1$.

Crossover and fitness value

For the crossover operation we implemented the so called one-point crossover: we select at random two parents from the present population and a crossover point from the numbers $\{1, \ldots, n-1\}$. A new solution is produced by combining the pieces of the parents. For instance, suppose we selected parents $(u_1, u_2, u_3, u_4, u_5)$ and $(v_1, v_2, v_3, v_4, v_5)$ and crossover point 2. The child solutions are $(u_1, u_2, v_3, v_4, v_5)$ and $(v_1, v_2, u_3, u_4, u_5)$.

To each individual we assign a fitness value. If it corresponds to a feasible bisection, i.e., the total node weight in both clusters stays within the limits l_τ and u_τ, we take the inverse of the objective function value in the incidence vector of the corresponding bisection cut. Otherwise we take a negated feasibility violation, i.e., the weight of the cluster which is greater than u_τ. In this way we obtain that the higher the fitness value the better the solution objective value. The p fittest individuals are selected from the expanded population and the next generation is created. The fitness value of the best individual is stored. It defines the fitness of the generation. We consider two stopping criteria of the genetic process. One is defined by f, called fitness loop number. It gives the limit of loops we perform without increase in the generation fitness. The second limit is the maximal number of loops we perform in one heuristic round. The output is the fittest individual. If it corresponds to a feasible bisection cut the inverse of its fitness value gives an upper bound for the MGBP.

In the first version we designed our algorithm to use the LP-relaxation of the MGBP. However, the solution of the linear relaxation can be easily replaced by an adequate solution of the semidefinite or another relaxation to the MGBP.

In Section 7.7 we computationally establish the parameters population size p, population growth k and mutation type $\mathcal{M}$. We also show the computational success of the hybridization as well as the overall performance of the algorithm.

6.3   Greedy Randomized Adaptive Search Procedure

The greedy randomized adaptive search procedure (GRASP) is a metaheuristic algorithm commonly applied to combinatorial optimization problems. GRASP typically consists of iterations made up from successive constructions of a greedy randomized solution and subsequent iterative improvements through a local search. The greedy randomized solutions are generated by adding elements to the problem's solution set from a list of elements ranked by a greedy function according to the quality of the solution they can achieve. To obtain variability in the candidate set of greedy solutions, well-ranked candidate elements are often placed in a restricted candidate list and chosen at random when building up the solution. GRASP was first introduced by Feo and Resende in 1989 [29].

Our implementation of GRASP follows the approach of Elrod at al. [28] for solving the 2-partition problem. This heuristic is based on exchanging nodes and can be seen as a randomized extension of the Kernighan-Lin algorithm. The searching process for a good solution is divided into two phases. The construction phase generates a pre-optimized feasible solution which is the starting point for further operations in the following improving phase. In the construction phase the initial solution is built by considering each node of the graph and assigning it to one of the two sets, say A and B, of the partition. The decision to which subset to allocate the node is supported by a greedy function that minimizes the augmented cut cost. Simultaneously the current node weights of sets A and B are compared. Since in our test setting the clusters should have almost equal node weights, the one with smaller weight is prioritized in the next move. Once all nodes are assigned to the subsets of the partition, the solution it is checked for feasibility. If one of the subsets is overloaded, a pairwise exchange is performed in order to establish the feasibility. The improving phase starts from the pre-optimized bisection generated by the construction phase. The solution is improved by exchanges of nodes impending the feasibility and promising decrease in the objective function.

Algorithm 6.2 GRASP's Construction Phase

Input Graph G the upper bisection bound u_τ, and parameter k.

1: $A \leftarrow v := rand(V)$; $\quad U := V \setminus \{v\}$; $\quad B := \emptyset$;

2: update cost values c_v, $v \in U$;

3: **repeat**

4:   **if** $f(A) \le f(B)$ **then**

5:     create candidate list K of k nodes with largest c_v, $v \in U$;

6:     $A \leftarrow v := rand(K)$; $\quad U \to v$;

7:     update cost values c_v, $v \in U$;

8:   **else**

9:     create candidate list K of k nodes with smallest c_v, $v \in U$;

10:     $B \leftarrow v := rand(K)$; $\quad U \to v$;

11:     update cost values c_v, $v \in U$;

12:   **end if**

13: **until** $U = \emptyset$

14: update exchange cost values c_v, $v \in V$;

15: **if** $f(A) > u_\tau$ (i.e., (A, B) is not a bisection) **then**

16:   create candidate list K with nodes in A sorted by increasing values c_v, $v \in A$;

17:   **while** $f(A) > u_\tau$ **do**

18:     take and remove the first element, say v, from K;

19:     $A \to v$; $\quad B \leftarrow v$;

20:   **end while**

21:   **if** $f(B) > u_\tau$ **then** randomly exchange nodes till $f(A)$, $f(B) \le u_\tau$ **end if**;

22:   update exchange cost values c_v, $v \in V$;

23: **end if**

24: **if** $f(B) > u_\tau$ **then**

25:   perform Steps 16-21 with respect to set B (in Step 21 sort K decreasingly);

26: **end if**

27: calculate cut weight $\omega_I := w(\Delta(A))$ of bisection (A, B);

28: update exchange cost values c_v, $v \in V$;

Output Initial feasible bisection (A, B), its edge cost ω_I, and exchange cost vector c.

Before we go into implementation details we introduce some notation. For a set of indices I we denote by $rand(I)$ a randomly chosen element from I. The greedy function evaluating the assignment of nodes is defined by

$$c_i = \sum_{j \in A} w_{ij} - \sum_{j \in B} w_{ij}, \quad i \in V,$$

where w is the edge cost vector. U is the set of vertices currently not in the partition $A \cup B$.

The Construction Phase

The construction phase is summarized in Algorithm 6.2. The first node v is selected randomly from V and assigned to the set A (Step 1). For all remaining nodes in $U := V \backslash \{v\}$ the greedy function $c(\cdot)$ is evaluated (Step 2). In each following iteration (Steps 4 - 12) a new node is added to the set with the currently smaller node weight. If it is $A's$ turn a candidate list with k nodes of largest values $c(\cdot)$ is established and a node is randomly selected from this list. Otherwise, a candidate list with k smallest values of the greedy function c is formed and a randomly selected node goes to B. After each move the greedy function is updated with respect to the nodes in U. Once U is empty we check if the so obtained partition is a feasible bisection. If it is not, i.e., say

$$f(A) = \sum_{i \in A} f_i > u_\tau,$$

we sort the nodes in B with respect to the increasing values of the greedy function c which previously is recalculated for each node. The nodes minimizing $c(\cdot)$ are moved to B till $f(A) \leq u_\tau$ (Steps 15 - 20). If finally $f(B) > u_\tau$ occurs we randomly swap nodes between A and B till feasibility is achieved[1]. The so obtained feasible bisection is handed over to the improving phase.

The Improving Phase

The pre-optimized solution produced by the construction phase undergoes a series of exchanges in order to improve the objective value. The approach of Elrod et al. [28] considers only pairwise exchanges which is satisfactionary for the 2-partition problem. However, in the bisection case the situation is more complicated, since nodes have different weights, and the cardinality of sets A and B do not have to be equal. For this reason we apply a more general exchange mechanism involving more than 2 nodes.

For convenience we define the **k-opt exchange**, which indicates the number of nodes involved in an exchange. If only one node is moved from A to B, or the other way round, then $k = 1$. If one node from A is swapped for a node from B we have a 2-opt exchange.

---

[1]We verified empirically, that this step is seldom necessary and, if needed, terminates quickly.

A difficulty associated by increasing k is the growing complexity and thus the running time, in case we want to perform all possible exchanges at each level. To reduce the running time we fix the amount of time which is used for node swapping. This time limit is established for each instance individually. We do not constrain the time for 1-opt and 2-opt exchanges but we measure their running time and take it as a limit for exchanges with $k \geq 3$. In this way the time limit is coupled with the size of the instance.

In Algorithm 6.3 we summarize the improving phase. For ease of exposition we give the exchange time limit t_e as a known parameter. In practice we calculate it as explained above. The number of exchange levels we perform are limited by the parameter n_e.

Algorithm 6.3 GRASP's Improving Phase

Input Initial bisection (A, B), its edge cost ω_I, exchange cost values c_v, $v \in V$, exchange order n_e, exchange time limit t_e.

1: **for** $k = 1, \ldots, n_e$ **do**
2:    **while** $t_e > 0$ **do**
3:       count time of exchange $s_t := get\_time()$;
4:       select randomly k nodes from $A \cup B$, let U be the set with selected nodes;
5:       $\overline{A} := (A \setminus U) \cup (U \cap B)$; $\overline{B} := (B \setminus U) \cup (U \cap A)$;
6:       **if** $f(\overline{A}) \leq u_\tau$ **and** $f(\overline{B}) \leq u_\tau$ **and** $g(U) < 0$ **then**
7:          $A := \overline{A}$; $B := \overline{B}$;
8:          update cut cost $\omega_I := \omega_I + g(U)$;
9:          update cost values c_v, $v \in U$;
10:      **end if**
11:       $e_t := get\_time()$; $t_e := t_e - (e_t - s_t)$;
12:    **end while**
13: **end for**

Output The bisection (A, B) and its edge cost ω_I.

For each $k = 1, \ldots, n_e$ we repeat the following steps as long as the time limit is not exceeded. We select k nodes from V and using the costs c_v calculate the gain g of a possible exchange as follows. Let U be the set of selected nodes. Obviously if $U := \{u\}$ then

$$g(U) = \begin{cases} c_u, & u \in A, \\ -c_u, & u \in B . \end{cases}$$

If we swap two nodes from different sets, i.e., $U = \{u, v\}$, $u \in A$, $v \in B$, then

$$g(U) = c_u - c_v + 2w_{uv} .$$

Generally,

$$
\begin{aligned}
g(U) &= \sum_{v \in A \cap U} \left(\sum_{u \in A \setminus U} w_{vu} - \sum_{u \in B \setminus U} w_{vu} \right) + \sum_{v \in B \cap U} \left(\sum_{u \in B \setminus U} w_{vu} - \sum_{u \in A \setminus U} w_{vu} \right) \\
&= \sum_{v \in A \cap U} c_v - \sum_{v \in B \cap U} c_v - \sum_{u,v \in A \cap U} w_{uv} - \sum_{u,v \in B \cap U} w_{uv} ,
\end{aligned}
$$

i.e., we measure how the cut weight will change if we swap nodes in $A \cap U$ against nodes in $B \cap U$. A negative gain means the swap pays off and is performed provided it does not cause a node weight overflow neither in A nor in B. To calculate the gain $g(U)$ we cache and update the costs c_v, $v \in V$, initialized in the construction phase. If the nodes are exchanged then we need only to reevaluate those c coordinates which correspond to nodes that are directly adjacent to one of the nodes being exchanged.

This heuristic was developed within the bachelor thesis of Lin [65].

6.4 The Software Package METIS

One of the known software packages solving a wide range of graph partitioning problem is the library METIS which is a set of serial programs for partitioning graphs, partitioning finite element meshes, and producing fill reducing orderings for sparse matrices. The algorithms implemented in METIS are based on the multilevel recursive-bisection, multilevel k-partitioning, and multi-constraint partitioning schemes.

The software is being developed at the Department of Computer Science & Engineering at the University of Minnesota, under the supervision of George Karypis, and available at *http://glaros.dtc.umn.edu/gkhome/metis*. The ideas behind the implemented algorithms are described by Karypis and Kumar in [56, 57], as well as in the user manual available at the homepage of the software.

We use the method *METIS_WPartGraphRecursive*(G, w_1, w_2) which computes a partitioning of a weighted graph with prescribed fixed partition weights by minimizing the cut weight. For the time being the library does not contain a routine to handle partition with variable partition weights. To avoid this drawback, and using the fact that the heuristic is extremely fast, we calculate a partition for each possible weights allocation allowed by the bounds u_τ and l_τ and take that one minimizing the weighted edge cut value, see Algorithm 6.4. The values w_1, w_2 give the exact percentage of the node weight for each cluster (Step 5). Note that we benefit from the fact that the node weights of graphs we consider are integral.

Algorithm 6.4 Using METIS Library

Input Weighted graph G, bounds l_τ and u_τ.

1: $cutweight := \infty$;
2: initialize array v of the length $|V|$ to store the partition;
3: initialize array v_{best} of the length $|V|$ to store the best partition;
4: **for** $i = 1, \ldots, (u_\tau - l_\tau)$ **do**
5:      calculate capacities for the clusters $w_1 := \frac{u_\tau - i}{u_\tau + l_\tau}$ and $w_2 := 1 - w_1$;
6:      $v := METIS\_WPartGraphRecursive(G, w_1, w_2)$;
7:      calculate cut weight *newcutweight* corresponding to v;
8:      **if** *newcutweight* $<$ *cutweight* **then**
9:         $v_{best} := v$; $cutweight := newcutweight$;
10:     **end if**
11: **end for**

Output v_{best}

Chapter 7

Computational Results

In this chapter we apply the polyhedral results presented in Chapters 3, 4 and 5, as well as heuristic methods described in Chapter 6, for solving the minimum graph bisection problem on various graph instances. At the beginning we give a short overview on empirical results already reported in the literature concerning exact solutions to graph partitioning problems. Next, we explain the basic concepts of SCIP, the brach-and-cut framework we use. The integration and incorporation of both semidefinite and linear relaxations within one branch-and-cut framework demands a one-to-one correspondence between the semidefinite and linear programming models for the MGBP. In Sections 7.3 and 7.4 we give the adequate models and details on the implementation of each relaxation. In the remaining of the chapter we discuss results of various test sets, which are performed on instances described in Section 7.6. In Section 7.7 we deliver insight how we established the variety of parameters steering the primal heuristics as well as the comparison of the performance of the algorithms. Our main contribution concerning the software development lies in the improvement of relaxations to the MGBP by applying cutting plane algorithms. In Section 7.8 we show the impact of our separation routines, introduced in Chapters 3 and 4, applied to the linear and the semidefinite relaxation. In Section 7.9 we continue comparing the performance of these relaxations by considering the linear relaxation of two different models and well as an incorporation of both the semidefinite and the linear relaxation within one solution process. Furthermore, we investigate the impact on both relaxations by changing the values of the bisection ratio τ. In [44] Helmberg reported a positive computational impact from the extension of the edge set of the underlying graph in combination with the semidefinite relaxation. In Section 7.10 we transfer this approach to the linear relaxation. The tests discussed in Sections 7.7 - 7.10 are performed on real word instances originated from VLSI design and scientific computations. The overall leadership of the semidefinite approximation is apparent. Finally, in Section 7.11 we compound our investigation on random instances, for which the linear relaxation outperforms the semidefinite approach.

The software we apply is implemented in C/C++ and compiled with Gnu gcc-3.4.5 on a Linux Debian platform. Most of the test were performed on HP Compaq DC7100

Pentium IV 540, 3.2 GHz HT with 1 GB RAM. The tests presented in the last section
were executed on AMD 64X2 4800+, 2.4 GHz and 4 GB RAM.

For convenience we store most of the tables with detailed results in Appendix B.

7.1   Known Numerical Results for Graph Partitioning

Besides the variety of heuristics methods developed to tackle graph partitioning problems, there are also exact approaches to solve these problems. Some of them concern
different approximations, e.g. in form of relaxations, and apply them in a branch-and-
bound framework. Others base on polyhedral investigations, which are then integrated
in a branch-and-cut algorithm in form of separation algorithms. Below we give a short
overview of the literature concerning experiments with a linear and a semidefinite relaxation to solve various problems in the area of graph partitioning.

In 1993, de Souza [24] applied polyhedral studies on the equicut polytope. In the tests
three classes of problems were used: mesh problems, compiler and VLSI design problems.
A graph with 148 nodes and 265 edges was the largest instance solved. Brunetta, Conforti, and Rinaldi [14] published in 1997 further results on the application of a branch-
and-cut algorithm with a linear relaxation to graph equipartition problems. Among
others they separate for triangle, clique and cycle inequalities and solve instances based
on complete graphs up to 100 nodes. In 1998, Ferreira, Martin, de Souza, Weismantel, and Wolsey [31] presented numerical results by solving the NCGPP. They used a
branch-and-cut algorithm with an LP-solver. Several classes of valid inequalities for the
associated polytope were separated, see Section 2.2. The authors were able to solve
sparse instances coming from parallel computing and VLSI design with up to 300 nodes.
A recent contribution concerning the max-cut problem is due to Barahona and Ladányi
[10]. They run experiments for max-cut problems arising from the so called *Ising spin
glass model*, a model in statistical physics, which has been studied in the literature since
1995, as well as for instances which are transformations of 0-1 quadratic programming
problems. They integrated the volume algorithm in a branch-and-cut framework to solve
approximately the linear relaxation and compared it with the standard approach based
on the dual simplex method. They were able to tackle instances not solved before.

One of the first experiments with semidefinite programming was presented by Helmberg
and Rendl [46] in 1998. They computed approximations for the max-cut problem by
solving the semidefinite relaxation with the interior point method. The relaxation was
tightened by triangle, clique and hypermetric inequalities. Various instances with up to
100 nodes could be solved to optimality. Lisser and Rendl [66] compared in 2003 the
linear and the semidefinite relaxation of the k-partition problems, where the nodes of the
graph are divided into k sets of possibly equal cardinality, arising in telecommunication
networks. Both relaxations were derived from a quadratic programming model and

strengthened by triangle inequalities. By means of instances with up to 900 nodes they showed that the LP-based approach is preferable in the case of a partitioning into many clusters, while the SDP-based approach works better when the number of clusters is small. In 2004 Helmberg [44] applied the spectral bundle method to compute the bounds for the MGBP on large sparse graphs with up to 20000 nodes. The semidefinite relaxation was tightened with odd cycle inequalities. In 2006 Rendl, Rinaldi, and Wiegele presented in [77] a method for finding exact solutions of the max-cut problem. They used a semidefinite relaxation combined with triangle inequalities, which they solved with the bundle method. Their algorithm can solve virtually any instance with about 100 variables in a routine way. The most recent results on using semidefinite programming to a graph partitioning problem was given by Armbruster in [4] in 2007. He integrated the spectral bundle method of Helmberg [44] together with the cutting plane algorithm within a branch-and-cut framework. He could tackle instances arising in the VLSI design and scientific computations as well random graphs known in the literature not solved before.

7.2 SCIP as a Branch-And-Cut Framework

We solve the minimum graph bisection problem with a state of the art branch-and-cut algorithm which is a partially non-commercial software SCIP for solving mixed integer programs. SCIP, the acronym for **S**olving **C**onstraint **I**nteger **P**rograms, is being developed at Zuse Institute Berlin and avaiable at *http://scip.zib.de*. SCIP in combination with CPLEX as LP-solver (*http://www.cplex.com*) offers one of the fastest currently available MIP-solver. It is a successor of SIP (Solving Integer Programs) written by Martin [70] in 1999. SCIP uses many ideas of SIP concerning solving mixed integer programming problems. Since the beginning of the development of SCIP in 2002 many features have been added improving the performance and utility of the tool. The upgrade to a constraint programming solver made it possible to handle non-linear constraints. This opened the door to incorporate solvers for non-linear relaxations, like a semidefinite solver in our case. Currently it can be used as a branch-and-cut framework for non-linear integer programming problems.

SCIP generally follows the branch-and-cut strategy explained in Section A.3. In Figure 7.1 we outline the operational stages in the solution process. For full implementation details we refer the reader to the PhD thesis of Achterberg [1], who carried out and supervised the development of SCIP from 2002 to 2007.

In the original version SCIP was designed to use exclusively the linear relaxation and the key role in the whole structure played **LP**, the linear programming relaxation of the input problem. To allow SCIP to handle other relaxations a **relaxator** plug-in has been developed. This tool provides all information needed for an external solver to create the corresponding relaxation with respect to the currently handled node of the branch-and-bound tree (b&b tree for short), and also makes it possible to transform back into SCIP the information of the external relaxator, like upper and lower bounds on the objective function and corresponding solutions. In Section 7.4 we give more details on how this tool is used to integrate the software solving semidefinite programs.

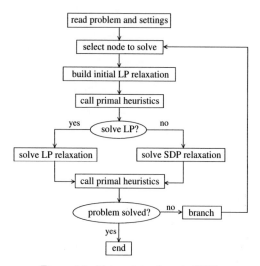

Figure 7.1: Main solving loop in SCIP.

We briefly describe the main steps SCIP performs to solve the MGBP using semidefinite or linear relaxation. The input file describes a graph for which we build up the linear integer programming model (IP3). SCIP derives the initial linear programing relaxation, which we can strengthen by adding triangle inequalities, see Section 7.3. Note that, since the SDP-relaxation uses LP-rows (see Section 7.4) this step is performed irrespective of which relaxation is solved. Now, it is possible to call heuristics, which find feasible solutions and deliver upper bounds. After that the relaxation of the problems is solved. Which solver, linear or semidefinite, is called first is decided by the priority parameter. The *frequency* parameter decides if a given solver is called at all and how often. In Sections 7.3 and 7.4 we describe how the cutting plane algorithms are handled by both relaxations. Once the cutting plane phase is finished, primal heuristics are called. If an optimal solution is not found, the dividing phase begins. Applying one of the so called **branching rules** SCIP splits the problem into subproblems and the solving loop continues till an optimal solution is achieved or another termination criteria are fulfilled like a time limit or a number of processed nodes of the b&b tree. For an overview on branching rules implemented in SCIP and their performance we refer to Achterberg, Koch, and Martin [2].

7.3 Linear Relaxation

In our computational test we applied the two linear integer models (IP2) and (IP3) for the MGBP and the corresponding linear relaxations. For convenience we cite these models from Section 2.3.

$$\min \sum_{e \in E} w_e y_e,$$

$$
\begin{array}{lrll}
\text{s.t.} & z_i - z_j & \leq & y_{ij}, & \forall ij \in E, \\
& z_j - z_i & \leq & y_{ij}, & \forall ij \in E, \\
& z_j + z_i & \geq & y_{ij}, & \forall ij \in E, \\
\text{(IP2)} & z_i + z_j & \leq & 2 - y_{ij}, & \forall ij \in E, \\
& l_\tau & \leq & \sum_{i \in V} f_i z_i \leq u_\tau, & \\
& y_{ij} & \in & \{0,1\}, & \forall ij \in E, \\
& z_i & \in & \{0,1\}, & \forall i \in V.
\end{array}
$$

In this formulation the z-variables force the values of y-variables so that the resulting vector corresponds to an incidence vector of a cut.

$$\min \sum_{e \in E} w_e y_e,$$

$$\text{s.t.} \quad l_\tau \leq \sum_{i \in V \setminus \{s\}} f_i y_{is} \leq u_\tau, \tag{7.1}$$

(IP3)

$$\sum_{e \in D} y_e - \sum_{e \in C \setminus D} y_e \leq |D| - 1, \quad \forall C \subseteq E, \ D \subseteq C, \ |D| \text{ is odd}, \tag{7.2}$$

$$y_e \in \{0,1\}, \ \forall e \in E.$$

In the above model the role of z-variable resumes the odd cycle inequality class which takes care that each solution to the problem is a cut in the graph G. There is a small drawback of this formulation. Namely, the number of all possible odd cycle constraints (7.2) is exponential in $|E|$. Nevertheless, we dispose of a polynomial time separation algorithm for these inequalities, see Algorithm 3.5. Given a vector $y \in \{0,1\}^{|E|}$, we can decide exactly and in polynomial time if it belongs to the set of feasible solutions of (IP3) or not. For this reason we do not include all the constraints (7.2) in the initial formulation of the problem, but require that each integer solution to the incomplete problem delivered during the branch-and-cut solution process satisfies all odd cycle constraints (7.2). This is checked by our odd cycle separation routine.

Integral solutions to the MGBP can be found in two different ways during the traversal of the b&b tree. On the one hand, the solution of the relaxation may be integral. On the other hand, a primal heuristic can find such a solution. As we already mentioned, the solutions are declared as feasible if the odd cycle separation routine does not find any violated inequality. If a solution of the relaxation is checked, the first odd cycle cut found is added to the LP. This cannot be done in the case of a solution of a heuristic, since the LP is not always available when primal heuristics are called. An execution of a heuristics is controlled by the following parameters.

(a) *priority* decides in which order the heuristics are called,

(b) *frequency* determines at which depth of the b&b tree the given method is called, and

(c) *timing* decides at which step of solving a node of the b&b tree the heuristic is called, e.g., before or after the relaxation solved.

The linear relaxations of the programs (IP2) and (IP3) arise by extending the domain of all variables from $\{0, 1\}$ to $[0, 1]$. Both relaxations can be strengthened by valid inequalities introduces in Chapters 3, 4 and 5. The inequalities are generated and added to the current LP-relaxation during separation rounds. The separation routines for each class of inequalities dispose of several parameters, which decides, when and how often a given separator is called and how many violated inequalities it produces.

(a) *priority* decides in which order the routines are called,

(b) if *delay* is on, the separator is only called when other, previously called, separators did not find any **cuts**, i.e., violated inequalities,

(c) *frequency* determines at which depth of the branch-and-bound tree the given routine is called,

(d) *number of rounds* determines how often the separator is called in one solving round of the relaxation,

(e) *number of cuts* gives the maximal number of cuts, which can be found during one call of the separator,

(f) *violation* is the minimal gap between the left and right hand side of the inequality allowing the cut to enter the LP.

Some of our cut separation algorithms are computationally expensive and produce cutting planes in a heuristic fashion. In this case, it is desirable to keep the retrieved cutting planes even if they are less useful for separating the current LP solution. The hope is that they might be applicable in later separation rounds or on other subproblems in the search tree. It is expensive to generate them again, and due to the heuristic nature of our separation algorithms we even might fail to find them again. Thus, we store the cutting planes for later use in a global **cut pool**.

In Figure 7.2 we scetched the cutting plane phase with the linear relaxation, which is executed at some node of the branch-and-bound tree. Initially the LP-relaxation is

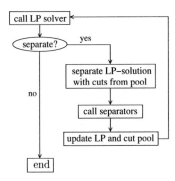

Figure 7.2: Cutting plane phase with LP-relaxation in SCIP.

solved and then the separation loop of an optimal LP-solution begins. A separation round is executed if the frequency of the cut pool or at least one separator equals the depth of the currently solved node of the branch-and-bound tree. The loop terminates if the maximal number of rounds of all separators exceeds the number of all executed rounds so far. At the beginning of each separation round the cut pool is checked for violated inequalities. Thereafter non-delayed separators active at the current node are called in the order of their priority. If no cut was found, the delayed separators have a new chance to generate inequalities. The LP and the cut pool are simultaneously updated with the cuts during the separation calls. Finally the LP is solved.

We solve the linear relaxation with the dual simplex algorithm, see e.g. Vanderbei [84], implemented within the software ILOG CPLEX 9.130.

7.4 Semidefinite Relaxation

We recall first the semidefinite programming model for the MGBP introduced in Section 2.4:

$$\min \langle C, X \rangle,$$

(SP)
$$\text{s.t.} \quad \langle f f^T, X \rangle \leq (u_\tau - l_\tau)^2,$$

$$\text{diag}(X) = e,$$

$$X \succeq 0.$$

To obtain the dual to (SP) we follow Section A.4. The constraint $\text{diag}(X) = e$ can be written as

$$\begin{pmatrix} \langle e_1 e_1^T, X \rangle \\ \vdots \\ \langle e_n e_n^T, X \rangle \end{pmatrix} = \begin{pmatrix} 1 \\ \vdots \\ 1 \end{pmatrix}.$$

By (A.3) the adjoint operator corresponding to the matrices $e_1 e_1^T, \ldots, e_n e_n^T$ fulfills

$$\mathcal{A}^T y = \sum_{i=1}^{n} y_i \, e_i e_i^T = \mathrm{Diag}(y).$$

Hence the dual to the problem (SP) reads

$$\max \langle e, y \rangle - (u_\tau - l_\tau)^2 p,$$

(DSP) $\qquad$ s.t. $\quad \mathrm{Diag}(y) - f f^T p + Z = C,$

$$p \geq 0,$$

$$Z \succeq 0,$$

$$y \in \mathbb{R}^n.$$

We use the spectral bundle method, explained in Section A.4, to tackle (DSP) in its equivalent form as an eigenvalue optimization problem,

$$- \min_{y \in \mathbb{R}^{|V|}, p \geq 0} |V| \lambda_{\max}(-C + \mathrm{Diag}(y) - f f^T p) - \langle e, y \rangle + (u_\tau - l_\tau)^2 p , \qquad (7.3)$$

see Proposition A.15. Note that the assumption of Proposition A.15 is fulfilled due to

$$I = \mathrm{Diag}(y) - f f^T p$$

for $y = e$ and $p = 0$. Furthermore, since $\mathrm{diag}(X) = e$ and $X \in S_n$ we obtain $\mathrm{tr}(X) = |V|$ yielding the correct multiplier a in the representation (A.5).

Separation and Support Extension

The spectral bundle method delivers approximations corresponding to feasible solutions of (7.3). Let such an approximation be given by a symmetric matrix $\overline{X}$. To be able to separate for the inequalities presented in Chapters 4, 5 and 3 we apply transformation (2.13)

$$\overline{y}_{ij} = \frac{1 - \overline{X}_{ij}}{2}$$

for all $ij \in E$. New valid inequalities for P_B in the form

$$\sum_{ij \in E} k_{ij} y_{ij} \leq k_0$$

are translated backwards into matrix constraints

$$\langle K, X \rangle \le \bar{k}_0,$$

where $K_{ij} = -\frac{1}{2}k_{ij}$ for all $ij \in E$ and

$$\bar{k}_0 = 2k_0 - \sum_{ij \in E} k_{ij},$$

and added to the primal semidefinite relaxation.

Due to the fact that any $X_{ij} \in \{-1, 1\}$ with $i \ne j$ describes the relation between the nodes i and j regardless whether $ij \in E$ or not, considering variables X_{ij} with $ij \notin E$ may bring some advantage. However, extending the underlying graph G to a complete graph appears computationally inefficient. In [44] Helmberg applies the separation for odd cycle inequalities to find out which new edges might be taken into consideration and shows the computational success of this approach. Once an edge is added to the graph and simultaneously the corresponding variable is added to the problem it can be used in separation routines for other classes of inequalities.

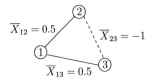

Figure 7.3: Due to the values $\overline{X}_{12}, \overline{X}_{13}$, and $\overline{X}_{13}$ adding the edge $\{2, 3\}$ one obtains a violated odd cycle inequality.

We demonstrate how the support extension is performed using odd cycle inequalities on the following example.

Example 7.1 Consider a subgraph $(V, E) := (\{1, 2, 3\}, \{\{1, 2\}, \{1, 3\}\})$ and assume that the separated solution of the semidefinite relaxation delivers values $\overline{X}_{12} = \overline{X}_{13} = 0.5$ and $\overline{X}_{23} = -1$, see Figure 7.3. Adding the edge $\{2, 3\}$ to the graph we obtain a violated odd cycle inequality

$$X_{12} + X_{13} \le 1 + X_{23},$$

or expressed in the y-variables,

$$y_{23} - y_{12} - y_{13} \le 0.$$

In this way the values of $\overline{X}_{12}$ and $\overline{X}_{13}$ are forced to change.

Armbruster developed a few more heuristics for support extension. They are described in [4], Section 4.3.4.

Having a solution of the semidefinite relaxation due to transformation (2.13) we immediately obtain y-variables, corresponding to edges in the graph G, used in each linear
integer programming formulation of the MGBP introduced in Section 2.3. Since only
model (IP3) is built exclusively on these variables, it gives us the desired one-to-one
correspondence with semidefinite model (SP). Therefore in our computational investigations if we apply both relaxations to solve an instance of the MGBP we consider model
(IP3).

SDP-Solver within SCIP's Framework

To solve the semidefinite relaxation we use the spectral bundle method developed by
Helmberg [43]. We summarize the method in Section A.4.3 and Algorithm A.2. The
integration of the SDP-solver into SCIP was developed by Armbruster and all implementation details are widely explained in his PhD thesis [4]. We only outline the main
solving loop of the SDP-solver drafted in Figure 7.4 below.

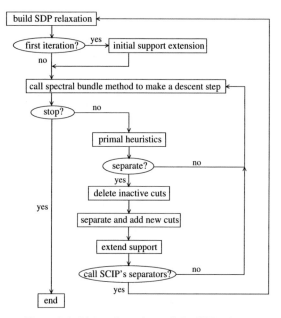

Figure 7.4: Main solving loop of the SDP-solver.

When the SDP-relaxation is called, first the (SP) model is created and supplemented
with the branching constraint representing the fixed variables in the considered subprob-

lem. Furthermore, the bisection constraint (7.1) from the (IP3) model and all additional inequalities from the current LP-relaxation are added as long as they are not recognized as inactive by an ancestor of the current branch-and-bound node.

In the first call of the SDP-solver the support is extended before the main loop is entered. After each descent step of the spectral bundle method one decides if the loop should be continued. This decision depends among others on the time already spent for computations and the progress of the dual bound with respect to the invested time. When the solution process continues primal heuristics and separators are called optionally prior to the next descent step of the spectral bundle. In the setting we use the primal heuristics are called every ninth iteration. A separation round is called if the current cutting plane surface model (see e.g. (A.8)) is sufficiently accurate (see Armbruster [4]). Before new inequalities are added the inactive ones are sorted out from the current SDP-relaxation. First the separators integrated in the SDP-solver are called. Afterwards, optionally, heuristics for support extension are executed. Finally, and also optionally, SCIP's separation routines are called, which obviously work on the extended support. This is achieved by adding new variables with zero objective values to the existing model.

Once the loop is finished the information useful for SCIP's b&b tree is shared. These are primal solutions, the final dual bound, new variables if not added yet and the branching decision. If the SDP-solver finished the solving phase of some node of the branch-and-bound tree its child nodes are created according to the solution of the SDP-relaxation. For that purpose Armbruster implemented several branching rules, which are described in [4] in Section 4.4.

7.5 Global Settings

In all our tests we took over general settings of SCIP concerning branching and node selection rules. We switched off all SCIP's standard primal heuristics and separators. Running preliminary tests we observed that they do not improve the efficiency of the solution process. Moreover, separation routines generating e.g. Gomory cutting planes cannot be applied for a solution of the semidefinite relaxation, since they base on the simplex algorithm.

In the majority of the tests we compute the minimal bisection cut for $\tau = 0.05$. For this value of τ we establish the best setting concerning primal heuristics and separators, see Sections 7.7 and 7.8. We also experiment with different values of τ. In Section 7.9 we present test results for $\tau \in \{0.01, 0.1, 0.3, 0.5\}$.

Since the parameters vary depending on which feature we test, we give a detailed description of the applied settings in the corresponding sections. In regard of the applied model of the MGBP and the relaxation we consider the following five combinations.

LP(IP2)   the model (IP2) with the linear relaxation;

LP(IP3)   the model (IP3) with the linear relaxation;

LP(IP3)+T   the model (IP3) together with all triangle inequalities and the linear relaxation;

LP(IP3)+SDP   the model (IP3) with the linear and the semidefinite relaxation, the semidefinite relaxation is only called in the root node of the b&b tree prior to the linear relaxation;

SDP   the model (IP3) with the semidefinite relaxation.

Concerning the parameters of the semidefinite relaxation used for the models SDP and LP(IP3)+SDP we took over the setting used in the final computations by Armbruster in [4] described there in Section 5.2 and 5.3. In this setting only the odd cycle separator is used. All other separators are switched off.

7.6   Graph Instances

In our investigations we consider three sets of graph instances. Apart from the origin of the graphs we give a description by using the following characteristic values:

(a)   the number of nodes and edges;

(b)   the minimal, maximal, and average node weight;

(c)   the minimal, maximal, and average edge weight;

(d)   the maximal and average node degree;

(e)   the density of the graph, i.e., the ratio $\frac{2|E|}{|V|(|V|-1)}$;

(f)   the diameter of the graph, i.e., the largest distance between nodes with respect to the number of edges on the path joining the nodes;

(g)   the algebraic connectivity, which is the value of the second eigenvalue of the Laplacian matrix associated with the graph, see (A.1) in Section A.1.

The last four values give an impression how strong the graph is connected.

VLSI design

In [54] Jünger, Martin, Reinelt, and Weismantel consider the placement problem in the layout design of electronic circuits, which consists of finding a non-overlapping assignment of rectangular cells to positions on the chip so that wireability is guaranteed and certain technical constraints are met. They develop approximation algorithms for the involved optimization problem using a graph theoretical formulation. Applying their code we obtain a partial placement for the chips *alue*, *alut*, *diw*, *dmxa*, *gap* and *taq*, which we use as input for the minimum graph bisection problem. The graph instances are described in Table B.1.

Nested Bisection

The second set of graphs was considered by Helmberg in [44]. They originate from nested bisection approaches for solving sparse symmetric linear systems, such as KKT-systems. The instances were handed over to Helmberg by Boeing as the standard bisection algorithms did not produced acceptable solutions as well as no approximation methods were known at that time. The graphs have equally weighted nodes and edges. Their names start with *kkt*. In the tables with computational results we refer to the graphs by the number of nodes and edges. The characterization of these graph instances is given in Table B.2.

Random Trees

In Section 7.11 we present test results performed on randomly generated graphs. For convenience we give the details on these instances at the beginning of the concerned section.

7.7 Primal Heuristics

In Chapter 6 we described heuristic methods, which we apply to acquire upper bounds for the MGBP. The design of the algorithms involves fine tuning of many parameters. We present below selected steps of customization towards the final layout of each of the method. These tests are carried out by solving model (IP3) with the linear relaxation. At the end of the section we give a comparison on the performance of the algorithms.

Genetic Algorithm

To make the genetic heuristic presented in Section 6.2 applicable to the wide class of graph instances and ensure the best possible performance we need to adapt several parameters. Running some preliminary tests we fixed the following parameters

(a) the total number of generations $g = 300$,

(b) the total number of loops without fitness improvement $f = 60$,

(c) the mutation rate $m_r = 1\%$.

In Table B.3 we give an overview on how the heuristic performs depending on the population size p, the population growth k, and the type of applied mutation $\mathcal{M}$.

We provide the following statistics: number of heuristic rounds r, total heuristic's running time in seconds T_h, achieved upper bound b_U and lower bound b_L. In these preliminary studies we considered $\varepsilon = 0.01$ by creation of the initial population, see page 121. No hybridization methods were applied.

We established that small instances like *taq.170.424* and *taq.228.692* can be most efficiently solved applying just the standard mutation type (Mutation 1). In these cases the solution time is proportional to the parameters p and k. On the contrary, on bigger instances like *diw.681.3104* and *taq.1021.6356* the heuristic performs better with problem related mutation types (Mutation 2,3,4) and the bigger population growth giving wider choice of solutions. Furthermore, better solutions of the LP-relaxation lead to better upper bounds computed by our heuristic and to less time consumption. We refer also to [5], where these results are published.

In the next test phase we investigated the impact of the hybridization methods on the performance of the algorithm. Following our earlier observation that the good quality of fractional LP-solution can be a significant help for the algorithm to deliver good solutions we looked for further developments. Thus we integrated edge costs in selected subroutines of the algorithm as a kind of decision support. We show below that such a hybridization delivers a good alternative to the standard random approach. These test are carried out using the settings

$$p = 150 \quad \text{and} \quad k = 4,$$

as we seek for an improvement of the upper bound for big instances. In addition all mutation types are selected randomly with a uniform distribution.

In our investigations we search for the possibly best selection of the set F in the sense that our heuristic constantly improves upper bounds and hence considerably contributes to the efficient termination of the branch-and-cut algorithm, i.e., the closing of the optimality gap:

$$gap = \frac{b_U - w^T \overline{y}}{w^T \overline{y}},$$

where b_U is the lowest upper bound found so far, $w \in \mathbb{R}_+^{|E|}$ is the vector with coefficients of the objective function and $\overline{y}$ is the solution of the linear relaxation to (IP3).

To this end, we vary the following parameters and methods. At the beginning we just increase the value of ε from 0.01 to 0.03 and 0.05. The success of the methods (F1) - (F3), see page 122, depends on the adequate selection of the pair (ε, ρ). It turns out that too small and too large cardinalities of F deteriorate the solution quality. The parameters ε and ρ appear to be coupled in a way that if we reduce ε we need to increase ρ and vice versa. For our tests we selected the pairs (ε, ρ) from set

$$B_{\varepsilon,\rho} := \{(0.03, 0.2), (0.05, 0.1), (0.05, 0.2), (0.01, 0.4)\}.$$

Next, we investigate the impact of the updated mutations on the solution while $F_\rho = \emptyset$ and only ε varies. Then we integrate them together with all variations of F_ρ described on page 122.

Instance	ε	ρ	F	$\mathcal{M}$	b_U	$\mu(b_U)$	$\sigma(b_U)$
taq.334.3763	0.03	$\sim$0.2	$\mathcal{R}$	+	341	530	184
	0.01	$\sim$0.4	F3	-	341	505	231
	0.01	0.4	F2	-	341	477	142
	0.05	$\sim$0.2	F2	+	341	461	116
	0.01	$\sim$0.4	F1	-	341	445	100
diw.681.3104	0.05	-	-	+	1228	1768	357
	0.03	$\sim$0.2	F2	+	1219	1903	378
	0.05	$\sim$0.1	F1	-	1205	1552	253
	0.05	$\sim$0.1	F1	+	1199	1785	389
	0.05	$\sim$0.1	F1	+	1185	1744	297
taq.1021.5480	0.05	0.1	F3	+	3678	3768	302
	0.01	$\sim$0.4	F3	+	3641	4120	403
	0.05	-	-	+	3425	3986	405
	0.01	0.4	$\mathcal{R}$	+	3403	3957	438
	0.05	$\sim$0.2	F3	+	3392	3992	481

Table 7.1: Test results on the hybridization of set F and Mutations 3 and 4.

It turns out that the simultaneous application of both, the hybrid set F and the hybrid mutations, not always contributes to better solutions, in comparison to a separate one. However, improvements based on the underlying problem, i.e., either to the initial phase of the algorithm or to the mutations, still perform better than random approaches.

Instance	N	r	heur. aver. time
taq.334.3763	300	50	25 sec.
diw.681.3104	100	20	51 sec.
taq.1021.5480	30	7	82 sec.

Table 7.2: Statistics on computations presented in Table 7.1.

In Table 7.1 we present the parameter selections for each instance, which contribute to the best upper bounds and hence to the reduction of the optimality gap. ε and ρ control the cardinality of the subsets F_ε and F_ρ. A tilde symbol $\sim$ in front of a ρ value indicates that the corresponding number X_ρ is selected randomly from the interval $[0, \rho]$, otherwise $X_\rho = \rho$. A plus symbol $+$ in the column $\mathcal{M}$ means that hybrid versions of Mutations 3 and 4 are applied. With respect to these variation of settings we compare the best upper bound achieved b_U, the average upper bound $\mu(b_U)$ and the standard deviation $\sigma(b_U)$. Table 7.2 shows statistics on the computations presented in Table 7.1. To obtain for each parameter selection an equal number of called heuristic rounds per instance we set the number of processed nodes of the b&b tree (N) as the computation limit. This number corresponds approximately to 3600 CPU seconds of the solution time of each instance. Furthermore, we give the number of heuristic rounds (r) and the average execution time of one heuristic round.

None of the methods (F1) - (F3) appears to be generally applicable for all instances. However, selecting an appropriate one contributes to good results. The first instance is solved most efficiently with the parameter pair $(\varepsilon, \rho) = (0.01, 0.4)$ independently from the hybridization form of the set F. $(\varepsilon, \rho) = (0.05, 0.1)$ and the method (F1) seems to be the best parameter selection for the second instance, while $\varepsilon = 0.05$ and method (F3) are the best for the third instance.

Instance	N	b_U	b_L	gap	time
taq.334.3952	317	-	299	inf	3600
diw.681.3104	109	-	389	inf	3600
taq.1021.5480	42	-	469	inf	3600

Table 7.3: Branch-and-cut without the genetic algorithm.

Finally in Tables 7.3, 7.4, and 7.5 we give a comparison on how the branch-and-cut performs without the genetic algorithm as well as the impact of the hybridization on the solution process. Table 7.3 shows the results when applying the pure branch-and-cut without the genetic heuristic. In Table 7.4 we give the best results, when the heuristic is applied in the non-hybrid version, just by varying the cardinality of the set F_ε. Table 7.5 contains the best results for each instance achieved with a hybrid version. b_L is the best lower bound. As time limit we used 3600 CPU seconds.

Instance	ε	N	b_U	b_L	gap	time
taq.334.3952	0.05	337	341	311	9.6 %	3600
diw.681.3104	0.01	108	1288	390	230.2 %	3600
taq.1021.5480	0.03	32	3896	435	795.6 %	3600

Table 7.4: Branch-and-cut with the non-hybrid genetic algorithm and the corresponding best selection of ε.

It pays off to apply the hybridization of the genetic heuristic with the exception of the instance *taq.334.3952*. But even in this case the hybrid version does not deteriorate considerably the result. For the other two larger instances the hybrid genetic algorithm yields the smallest gap by improving the upper bound. This keeps the branch-and-bound tree smaller so that the best lower bound also benefits.

Instance	N	b_U	b_L	gap	time
taq.334.3952	327	341	309	10.1 %	3600
diw.681.3104	115	1185	445	166.2 %	3600
taq.1021.5480	32	3403	450	656.2 %	3600

Table 7.5: Branch-and-cut with the hybrid genetic algorithm.

All computations presented in this section were executed on a 2.6 GHz Pentium IV processor with 2 GB main storage.

As one can read from the above fine tuning of the parameters is not easy to establish a universal settings, which best performs on all instances. Hence we adjusted partially basing on the intuition the final parameter setting for the genetic algorithm. Below we list parameter values used in computations presented in the subsequent sections:

number of generations	g	300
number of loops without improvements	f	15
population size	p	350 for $|V| < 500$, 300 for $500 \geq |V| \leq 1000$, 250 for $|V| > 1000$
population growth rate	k	2.2
creating initial population:	ε	randomly chosen from the interval $[0.01, 0.05]$
	X_ρ	randomly chosen from the interval $[0.2, 0.4]$
	F_ρ	methods (F1), (F2), (F3) and $\mathcal{R}$ all at random
mutation rate	m_r	1%
mutation type	$\mathcal{M}$	all at random, hybridization of 3 and 4 is performed randomly.

GRASP

The sensitive part of the heuristic based on the greedy randomized adaptive procedure is the improving phase. We refer to Section 6.3 for details. Once an initial solution is constructed we seek for a better one by swapping the nodes between the clusters of the corresponding bisection. Due to the fact that we consider graphs with not equally weighted nodes, we apply exchanges of more than one node. As one can expect, the more nodes are involved in a swap, the more possibilities are to check. Thus, on the one hand, more time is consumed, on the other hand, a better solution is obtained. In Figure 7.5 we plotted the dependency between the upper bounds achieved at k-opt exchange by increasing k, i.e., the number of swapped nodes, and the time needed for one run of the heuristic. We take average numbers of a sample of 9 $vlsi$-graphs for $k \in \{1, 2, 3, 4\}$. The detailed numbers can be found in Tables B.4 and B.5. To point out the improvement of the upper bounds with growing k we normalized the bounds with respect to the biggest one on the y-axis. The boxes and circles correspond to the exchanges with and without limited time, respectively.

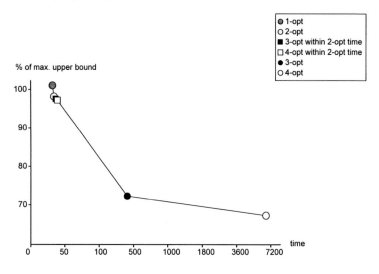

Figure 7.5: Performance of the GRASP-heuristics on a sample of $vlsi$-graphs.

As one can read from the chart, increasing k up to 4 yields a considerable improvement of the upper bound. However, one has to accept the disproportional increase in time. Unfortunately, reducing the time spent for a search of an exchange for bigger k does not pay off, as the boxes corresponding to 3-opt and 4-opt with the time limit show. Here, the time for swaps was fixed to the time spent on swapping in 2-opt.

On the basis of the above tests we decided to keep small the number of exchanged nodes, i.e., we set $k := 2$ in the final setting and do not limit the swapping time. We run the heuristic several times in hope that we casually obtain a good solution.

METIS versus GA and GRASP

Once we established the internal parameters of the heuristics we integrated them in the full solution process of SCIP. Due to the fact that the heuristics METIS, described in Section 6.4, uses exclusively the original formulation of the problem we run it only once at the beginning of the solution process, before the root node of the branch-and-bound tree is solved. Running preliminary tests we observed a good performance of the METIS software with respect to the time efficiency as well as the quality of delivered upper bounds for the MGBP in comparison with GRASP and the genetic algorithm. Therefore we redesigned the latter two methods as improving heuristics. We abandon occasionally (at some random decision) the constructing phase of GRASP and start its improving phase with the best solution known so far. GRASP is designed to solve the MGBP without supporting information in form of current solutions of relaxations. However, since in the algorithm random decision are taken, it can come up with different solutions in a subsequent call. In Tables B.4 and B.5 in the last line for each instance we give some insight how GRAPS performs as an improving heuristic. For nearly the half of tested graphs GRASP obtained a better upper bound than that of METIS and in a reasonable amount of time. Nevertheless, GRASP works much worse than the genetic algorithm. Therefore it is executed with the smallest priority and the frequency 30 in the final setting.

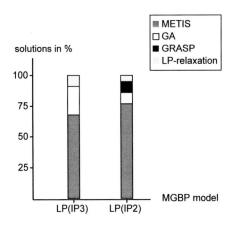

Figure 7.6: Performance of the primal heuristics on *vlsi*- and *kkt*-graphs.

To make out of the genetic heuristic an improving method we add to the initial population up to ten primal solutions found previously, see Algorithm 6.1. This is done at random decision, about every second call. Being the most time consuming method it is called with frequency 20 in the final setting.

We apply the above settings on the primal methods to the tests that follow. In Figure 7.6 we give the statistic on how often each heuristic found the best upper bound while solving all *vlsi*- and *kkt*-instances by applying the linear relaxation to the models (IP2) and (IP3). In case of GRASP and the genetic algorithm the number indicates how often each heuristic managed to improve METIS and the other one. The results are based on computations presented in Tables B.6 - B.8. In the column with head h we marked which heuristic found the best upper bound. "m" stands for METIS, "g" for the genetic algorithm, "G" for GRASP, and "r" if the solution was found by a relaxation[1]. The majority of the best solutions is delivered by METIS. The genetic heuristic manages sometimes to improve METIS and works better with solutions of the linear relaxation to (IP3) than with the model LP(IP2). In the latter case it cannot even improve GRASP's upper bounds.

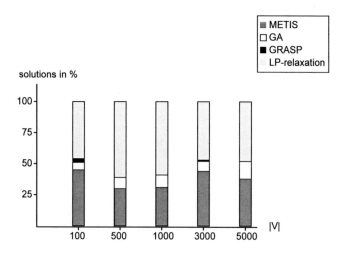

Figure 7.7: Performance of the primal heuristics on random trees.

In the second chart, Figure 7.7, we give the percentage of best solutions found by each heuristic while solving the random instances with the models LP(IP2), LP(IP3) and LP(IP3)+T. The statistic is made up on the basis of computations presented in Section 7.11. The outcome is similar as above. The genetic algorithm finds the best solution more frequently than GRASP, but both fail quite often to improve METIS.

---

[1]However, if SDP-solver delivered the upper bound then it is due to its internal primal heuristic.

7.8   Separation

In this section we apply our theoretical investigations concerning tightening the relaxations of the MGBP with valid inequalities for the bisection cut polytope P_B in the branch-and-cut method. Here, we integrate in SCIP the separation algorithms presented in Chapters 3, 4, and 5. The efficiency of the inequalities is tested independently with the linear and the semidefinite relaxation. Since we generally want to measure the impact on the relaxations, we solve only the initial relaxations, in the root node of the b&b tree.

Each relaxation is solved repeatedly as long as a given separation routine finds violated inequalities. In one separation round multiple cuts can be found. This is due to the fact that we repeatedly start to build the supporting graph of the inequality from a randomly selected node. Hence we add as many cuts to the relaxation as we managed to find. The results are presented in Tables 7.6 - 7.9. We set the time limit to 4 CPU hours.

Each column contains the best lower bound per instance achieved by separating the given inequalities. By testing the linear relaxation we consider the model (IP3). Hence the lower bound achieved by the separation for only odd cycle inequalities can be seen as the bound given by the "pure" relaxation. Since the odd cycle separator in some sense improves the feasibility of the initial problem, it is not switched off when other separators are tested.

In the heads of the tables we make use of the following shortcuts:
> *oc*   odd cycle inequalities (by the linear relaxation corresponds to the pure relaxation);
> *ccwt*   complementarity cycles with tails inequalities;
> *ct*   complementarity tree inequalities;
> *cycle*   cycle inequalities;
> *bkw*   bisection knapsack walk inequalities;
> *kt*   knapsack walk inequalities;
> *all*   all inequalities are applied;
> *ccc*   cycle, complementarity cycles with tails inequalities and complementarity tree inequalities;
> *sdp*   pure semidefinite relaxation;

We marked bold and italic the best and second best upper bound per line, respectively. In Figures 7.8 - 7.11 we plotted the bounds achieved by each separation routine relative to the best known lower bound, in many cases equal to the optimal solution. We use the results published by Armbruster in [4].

Applying the linear relaxation the knapsack tree inequalities have the biggest impact on the improvement of the lower bound by solving the MGBP, as one can read from Tables 7.6 and 7.7. Nevertheless, other inequalities also strengthen the relaxation, because if we apply all inequalities the bound increases sometimes considerably as, e.g., for the instances *diw681.3104* or *alut2292.494500*.

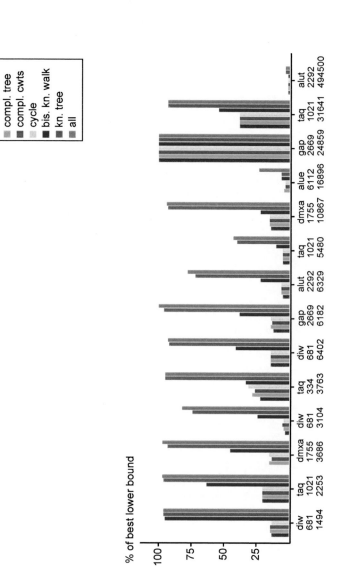

Figure 7.8: Performance of separation routines with LP-relaxation applied to *vlsi*-graphs.

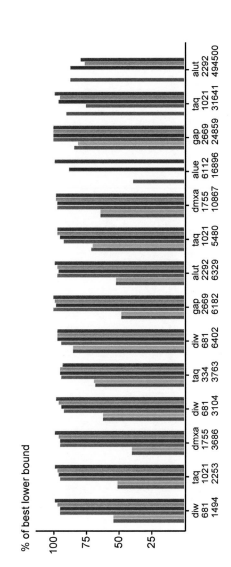

Figure 7.9: Performance of separation routines with SDP-relaxation applied to *vlsi*-graphs.

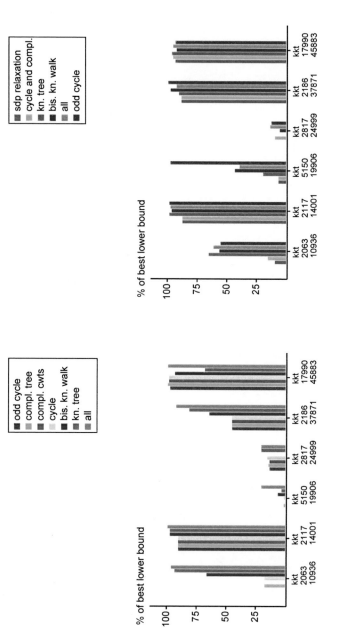

Figure 7.11: Performance of separation routines with SDP-relaxation applied to *kkt*-graphs.

Figure 7.10: Performance of separation routines with LP-relaxation applied to *kkt*-graphs.

problem	oc	ccwt	ct	cycle	bkw	kt	all
diw681.1494	18.4	19.4	19.4	18.9	134.9	**136.3**	135.9
taq1021.2253	23.1	24.1	24.1	23.6	74.1	*113.2*	**113.9**
dmxa1755.3686	0.0	12.4	13.9	13.8	42.6	*87.1*	**91.1**
diw681.3104	34.8	47.2	39.4	45.4	238.4	*744.4*	**829.2**
taq334.3763	75.5	88.2	95.7	106.6	111.8	*324.8*	**324.9**
diw681.6402	46.8	47.7	47.8	47.4	136.2	*304.6*	**306.3**
gap2669.6182	8.6	9.6	10.1	10.1	28.3	*71.1*	**73.7**
alut2292.6329	3.8	4.3	4.3	4.3	17.1	*55.3*	**59.8**
taq1021.5480	74.1	79.9	78.2	75.7	154.4	*639.8*	**689.9**
dmxa1755.10867	20.4	21.4	21.4	21.4	31.7	*137.1*	**138.6**
alue6112.16896	0.0	3.8	4.8	4.5	*8.2*	7.8	**31.0**
gap2669.24859	55.0	55.0	55.0	55.0	55.0	55.0	55.0
taq1021.31641	151.8	153.0	152.7	152.2	215.1	*372.5*	**374.6**
alut2292.494500	559.0	559.7	559.0	559.3	740.3	*1571.3*	**1966.2**
mean lower bound	23	32	33	33	77	162	186
mean time	40	509	297	520	8389	11646	10267

Table 7.6: Strengthening the linear relaxation by cut separation applied to *vlsi*-graphs.

problem	oc	ccwt	ct	cycle	bkw	kt	all
kkt2063.10936	0.0	0.0	1.0	1.0	4.0	*5.6*	**5.8**
kkt2117.14001	509.2	509.7	509.7	509.7	548.1	*548.1*	**560.7**
kkt5150.19906	0.0	0.0	1.0	1.0	*8.8*	4.5	**30.4**
kkt2817.24999	6.2	6.2	7.1	7.1	7.1	**9.9**	*9.8*
kkt2186.37871	943.0	943.8	943.5	943.5	1337.5	*1673.1*	**1920.0**
kkt17990.45883	6382.7	6440.3	*6519.8*	6470.2	6156.5	4467.0	**6553.3**
mean lower bound	52	52	53	53	102	100	152
mean time	245	245	331	261	4922	8552	6477

Table 7.7: Strengthening the linear relaxation by cut separation applied to *kkt*-graphs.

problem	sdp	ccc	kt	bkw	all	oc
diw681.1494	77.3	77.3	134.7	135.1	*137.9*	**140.6**
taq1021.2253	60.1	60.1	112.4	112.9	*115.0*	**116.8**
dmxa1755.3686	37.5	37.6	89.0	89.3	*90.3*	**92.9**
diw681.3104	630.7	624.4	935.0	954.6	*969.4*	**988.8**
taq334.3763	234.0	234.7	320.5	**324.8**	*324.7*	317.9
diw681.6402	280.8	280.3	310.4	*320.7*	319.7	**322.7**
gap2669.6182	35.8	35.8	*74.0*	72.7	73.0	**74.0**
alut2292.6329	39.7	39.7	74.6	74.0	*74.9*	**76.2**
taq1021.5480	1122.0	1109.2	1469.9	1510.8	*1538.9*	**1540.6**
dmxa1755.10867	94.6	94.7	142.0	143.0	**143.7**	*143.5*
alue6112.16896	52.9	53.2	99.9	*117.6*	98.0	**135.1**
gap2669.24859	46.0	44.5	55.0	55.0	55.0	55.0
taq1021.31641	359.4	232.9	301.8	386.6	*396.7*	**398.6**
alut2292.494500	**53950**	16808.5	46404.7	*53374.5*	47006.8	51071.4
mean lower bound	184	163	270	282	279	289
mean time	120	1697	3386	5684	4002	1476

Table 7.8: Strengthening the semidefinite relaxation by cuts separation applied to *vlsi*-graphs.

problem	sdp	ccc	kt	bkw	all	oc
kkt2063.10936	0.5	0.9	**3.9**	3.3	*3.6*	3.2
kkt2117.14001	493.7	493.6	544.4	541.9	*551.5*	**556.7**
kkt5150.19906	9.7	9.5	28.9	*64.0*	57.9	**144.8**
kkt2817.24999	0.0	4.5	0.0	2.4	**6.5**	*5.8*
kkt2186.37871	1837.9	1832.1	1865.6	*2013.1*	1907.3	**2064.4**
kkt17990.45883	6102.9	6279.4	**6320.8**	6088.9	*6287.2*	6109.7
mean lower bound	55	77	95	123	144	164
mean time	412	2294	10535	7683	12171	5650

Table 7.9: Strengthening the semidefinite relaxation by cut separation applied to *kkt*-graphs.

Considering the pure relaxation of model (IP3), i.e., only odd cycle inequalities separator is called, one achieves very poor lower bounds. Only for *kkt17990.45883* the efficiency of the odd cycle algorithm is visible. Therefore the polyhedral investigations yielding new classes of valid inequalities for the bisection cut polytope P_B pay off when considering the linear relaxation of the basic model of the MGBP.

The situation looks quite differently when applying the semidefinite solver. Here, the relaxation of the basic model (SDP) yields already good lower bounds. For very large instances like *alut2292.494500* the separation routines only slow down the solution process and thus lead to worse bounds when the computing time is not long. The best results are achieved when exclusively odd cycle inequalities separator is applied. The cycle, complementarity tree, and complementarity cycles with trees inequalities turn out to be completely useless. The inequalities employing the knapsack condition, i.e., the knapsack tree and bisection knapsack walk, perform nearly the same and bring also good bounds but on higher time costs in comparison to odd cycle inequalities.

7.9   Linear versus Semidefinite Relaxation

In [4] Armbruster presented first results comparing the linear versus the semidefinite relaxation. He considers only the model (IP3) and (SDP), and each relaxation are applied separately. We continue the empirical investigations by involving the other linear integer model (IP2), as well as by combining both relaxations. Finally we vary the bisection ratio τ.

On the basis of our observations concerning the performance of the separation routines summarized in Section 7.8 we decided on the following settings. The meaning of each parameter was described in Section 7.3.

separator	*priority*	*frequency*	*nb. of cuts*[*]	*nb. of rounds*[*]	*delay*	*violation*
knapsack tree	1	1	150/50	no limit/1	false	0.1
odd cycle	2	1	200/50	no limit/1	false	0.1
bis. knapsack walk	3	10	50/50	100/1	true	0.0001
compl. tree	4	5	50/50	10/1	true	0.05
cycle	5	5	10/10	10/1	true	0.05
compl. cycl. & trees	6	5	50/50	10/1	true	0.05

[*] in root/ not in root node of the b&b tree

The settings for primal heuristics are described on page 148. The computation time for tests presented in this section was set to 10 CPU hours.

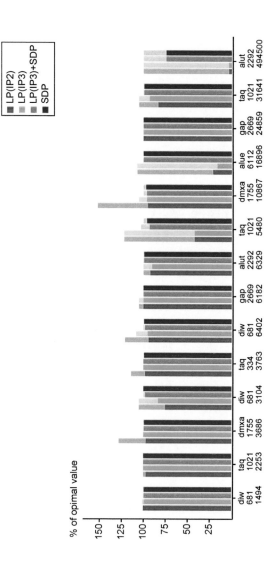

Figure 7.12: Lower and upper bounds calculated with *vlsi*-graphs by using various models for the MGBP.

In Figures 7.12 and 7.13 we depicted the upper and lower bounds achieved by solving the MGBP on *vlsi-* and *kkt*-graphs, respectively, with different models and relaxations. The dark bars correspond to lower bounds, the lightened bars to upper bounds.

The overall better performance of the semidefinite relaxation in comparison to the linear one was already reported in [4] and confirmed by our computations, see the yellow and blue bars in the charts. Applying the models LP(IP3) and LP(IP2), which exclusively use the linear relaxation, one obtains comparable results, although (IP2) model gives a complete formulation of the MGBP and does not need much more variables than (IP3). One observes a general weakness of primal methods delivering upper bounds.

The optimality gap decreases when we integrate the semidefinite relaxation. The aggregation of the primal solution performed within the spectral method (see Section A.4.3) as well as following improvement heuristics give good upper bounds already in the first calls of the semidefinite relaxation.

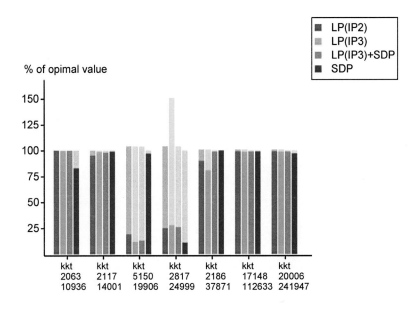

Figure 7.13: Lower and upper bounds calculated with *kkt*-graphs by using various models for the MGBP.

In our preliminary tests we diagnosed that alternating calls of linear and semidefinite relaxation are too time consuming. The LP-solver does not benefit from the lower bounds delivered by the SDP-solver and the latter one delivers anyway better results. Hence one of the solvers is actually redundant. This is why we decided to consider the model LP(IP3)+SDP, where the semidefinite relaxation is solved only in the root node of the b&b tree prior to the linear relaxation. As the corresponding gray bars show, in this way we could reduce the optimality gap.

The charts were created on the basis of the numerical results presented in Tables B.6, B.7, and B.8. Beyond the lower and upper bounds we give the computation times and number of solved nodes of the b&b tree. For an overview we calculated the geometric average of these values in Tables 7.10 and 7.11.

setting	time	gap	b&b nodes	upper bound	lower bound
LP(IP2)	32883	17	3	222	156
LP(IP3)	27144	8	12	208	156
LP(IP3)+SDP	**14123**	1	2	200	196
SDP	17278	**1**	42	**199**	**197**

Table 7.10: Geometric means of results given in Tables B.6 and B.7.

The LP(IP2) model needs the most time which is consumed by solving the LP-relaxation in the root node of the b&b tree. This explains the small number of solved b&b nodes. The inequalities, which cut off the solution of LP-relaxation, poorly improve the approximation. Nevertheless, the most of *kkt*-graphs are more efficiently solved by LP(IP2) than by LP(IP3).

settings	time	gap	b&b nodes	upper bound	lower bound
LP(IP2)	28026	10	4	462	288
LP(IP3)	**25143**	10	23	488	271
LP(IP3)+SDP	25634	6	5	461	279
SDP	36693	**5**	16	**456**	**321**

Table 7.11: Geometric means of results presented in Table B.8.

The combination of the linear and semidefinite relaxation in the setting LP(IP3)+SDP pays off the most when applied to the *kkt*-graphs. But also for *vlsi*-graphs it brings comparable bounds to SDP at a lower time cost. The low number of b&b nodes shows that the strengthening of the relaxation by many separation rounds in the root node of the b&b tree delivers a lower bound close to the optimal value, and that the primal solution delivered by the SDP-solver is of very good quality.

Varying τ

The good performance of the semidefinite relaxation is not sensitive to the value of the bisection ratio τ as results presented in this section show. Here we consider the standard models LP(IP3) and SDP for the linear and the semidefinite relaxation, respectively. The value of τ gives the maximal feasible difference between node capacities of the two bisection clusters. If τ is close to zero then both clusters should have the weighted node sum almost equal. The higher the value of τ the more flexibility one gains in assigning the nodes to the clusters.

In Tables 7.12 and 7.13 we summarized the numerical results concerning the performance of the LP- and SDP-relaxation for the τ-values 0.01, 0.05, 0.1, 0.3, and 0.5. We give the geometric average of the computation time, the optimality gap, the number of the b&b tree nodes, the upper bound, and the lower bound. The detailed numbers for the considered instances can be found in Appendix B. The linear relaxation pays off apparently only if we have nearly the equipartition case, i.e., $\tau = 0.01$ and *kkt*-graphs. For $\tau = 0.3$ and $\tau = 0.5$ a few instances are solved very quickly. However, for the other instances it LP-relaxation performs very badly, as the poor lower bounds show. SDP-relaxation delivers in average better lower and upper bounds. Only for $\tau = 0.5$ LP-relaxation's average lower bound is better. Nevertheless, for all considered values of the bisection ratio τ the upper bounds obtained by the linear relaxation are close to those of the semidefinite one.

τ	setting	time	gap	b&b nodes	upper bound	lower bound
0.01	LP(IP3)	26627	8	6	229	166
	SDP	**16058**	2	41	213	207
0.05	LP(IP3)	27144	8	12	208	156
	SDP	**17278**	1	42	199	197
0.1	LP(IP3)	**27895**	14	6	211	150
	SDP	35772	2	107	196	191
0.3	LP(IP3)	**25764**	16	6	188	123
	SDP	31162	5	111	175	165
0.5	LP(IP3)	**30775**	24	5	168	99
	SDP	32502	10	120	152	130

Table 7.12: Geometric means of results for *vlsi*-graphs by varying τ given in Table B.9 ($\tau = 0.01$), Tables B.7 and B.6 ($\tau = 0.05$), Table B.10 ($\tau = 0.1$), Table B.11 ($\tau = 0.3$), and Table B.12 ($\tau = 0.5$).

τ	setting	time	gap	b&b nodes	upper bound	lower bound
0.01	LP(IP3)	**29173**	**11**	9	539	287
	SDP	36249	13	9	533	316
0.05	LP(IP3)	**25143**	10	23	488	271
	SDP	36693	5	16	456	321
0.1	LP(IP3)	**23530**	5	26	333	231
	SDP	24906	3	7	329	294
0.3	LP(IP3)	15274	5	5	265	180
	SDP	**14954**	2	4	260	253
0.5	LP(IP3)	**12322**	9	4	226	**145**
	SDP	18469	8	7	218	128

Table 7.13: Geometric means of results for *kkt*-graphs by varying τ given in Table B.9 ($\tau = 0.01$), Table B.8 ($\tau = 0.05$), Table B.10 ($\tau = 0.1$), Table B.11 ($\tau = 0.3$), and Table B.12 ($\tau = 0.5$).

7.10   Support Extension in Linear Relaxation

In Section 7.4 we pointed out an enhancement of the semidefinite relaxation achieved by adding new, specifically constructed edges to the underlying graph and thus variables to formulations of the MGBP. This approach, called support extension, proved to be successful when applied by solving the semidefinite relaxation of the max-cut problem and the MGBP, see Helmberg [44] and Armbruster [4].

In this section we investigate the impact of such a method when applied to the linear relaxation. Unfortunately, the support extension does not yield the expected results. It slows down the LP-solver and separators, which now have to work on a denser graph. The new variables apparently do not stimulate the separation methods to produce sharper cuts.

We tried two approaches. In the first one, which we call *sup1*, we took over the extended support, which the semidefinite relaxation ended up in the root of the b&b tree with. We expected that those variables or edges, which turned up to be useful for the SDP-model, could also positively affect the linear model. Once an instance is read in SCIP and the initial problem is constructed, we add the new variables to the problem and start the solution process.

The second method, referred to as *sup2*, also uses the variables selected by solving the SDP-model. However, we apply a more sophisticated way to construct new variables for the linear program. We follow the idea of Köppe et al. presented in [37, 61], which paid off by solving certain knapsack problems. Due to the knapsack condition in the formulation of the MGBP we expected that this approach could also work in our case. Here, we use the fact that an edge corresponding to some new variable is a part of a triangle with two edges incident on the center of the star, see the derivation of the model (IP3) in Section 2.3. This triangle-relationship of the edges should be exploited by the corresponding variables.

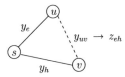

Figure 7.14: Adding aggregated variable z_{eh} instead of y_{uv}.

Let y_e and y_h be variables corresponding to edges e and h, already existing in the graph, incident on the center of the star $s \in V$. Let z_{eh} correspond to the new variable suggested by the SDP-solver, see Figure 7.14. We add the constraint

$$y_e + y_h + z_{eh} \leq 1$$

to the problem formulation. In this way we change the meaning of the variables y_e, y_f and demand that if $y_e = y_h = 1$, in the original sense, then $z_{eh} = 1$, otherwise $z_{eh} = 0$. The role of the original variables y_e and y_f is now taken by $y_e + z_{eh}$ and $y_h + z_{eh}$, respectively. Since the variable z_{ef}, in the sequel referred to as **aggregated variable**, substitutes in some sense two variables, we obtain a stronger formulation of the original problem.

New cuts added to the LP during separation routines are updated accordingly. If at least one of the original y-variables is in the cut than also the associated z-variable is added. The coefficient of z_{ef} depends on which y-variable is in the cut. If only one, say y_e, then z_{eh} is multiplied by the coefficient of y_e. If both y_e and y_h are in the cut then z_{eh} is multiplied by the sum of the coefficients of y_e, y_h.

We abandon a new variable z_{eh} if a z-variable associated either with y_e or y_h is already added. Too many z-variables corresponding to one y-variable weaken the cuts, as the following example shows.

Example 7.2 Suppose we found a valid inequality of the form

$$\sum_{\bar{e} \in \overline{E}} y_{\bar{e}} + y_e + y_h + y_g \geq 4, \tag{7.4}$$

where e, h, g are incident on the star center s and $\overline{E} \subset E$. Furthermore, assume that we want to add the new z-variables z_{eh}, z_{eg}, and z_{hg}. Then the above inequality is transformed to

$$\sum_{\bar{e} \in \overline{E}} y_{\bar{e}} + y_e + y_h + y_g + 2z_{eh} + 2z_{eg} + 2z_{hg} \geq 4. \tag{7.5}$$

Consider a solution $y \in \mathbb{R}^{|E|}$ with $y_e = y_h = y_g = 1$ and $y_{\bar{e}} = 0$ for $\bar{e} \in \overline{E}$. Then inequality (7.4) cuts off such a solution, while (7.5) does not.

Obviously, one could think of an more appropriate adjustment of the coefficients of the z-variables in such cases. But since the first tests did not show promising results, we abandoned further enhancements. Another obstacle we obtain by introducing aggregated variables of adjacent edges is of combinatorial nature. In the above example each of the variables z_{eh}, z_{eg}, z_{hg} could also be used as the aggregate of the other ones. It remains an open question, how to handle such a case.

In Table B.13 we put together the numerical results. We take our standard model LP(IP3) for solving the MGBP with the linear relaxation and extend the support by the methods *sup1* and *sup2*. The computation time limit was set to 5 hours. In the third column we give the number of variables considered depending on which method was applied. The neighboring column gives the percentage of the number of added variables. The support of the underlying graph is considerably extended by solving the MGBP with the semidefinite relaxation. We obtain in average 30% more variables. The number of the variables is sometimes even doubled. The selection of variables considered in the other approach varies around 2%.

In Figure 7.15 we plotted the lower bounds achieved by each method with respect to the best known lower bound. The large amount of new variables in *sup1* (black bars) drops considerably the performance. By applying *sup2* we stay close to the lower bound of LP(IP3) but do not manage to exceed it. Due to the failure of these experiments, we assert that extending the number of variables does not pay off in the linear programming relaxation or one has to resort to more sophisticated methods.

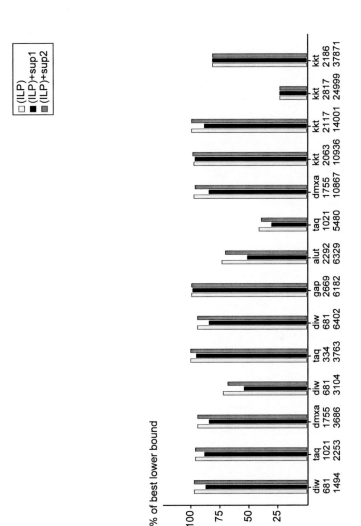

Figure 7.15: Comparison of lower bounds achieved by support extension in the model LP(IP3) applied to a sample of *vlsi*- and *kkt*-graphs.

7.11   Random Trees

The final tests are performed on random instances. The idea originates in the assumption that the efficiency of the semidefinite relaxation in comparison with the linear relaxation should be somehow coupled with the density of the graph. This is motivated by the fact that if we have a tree with n nodes then the linear formulation needs in worst case twice as much edge variables (due to the extension with the star), while in the semidefinite model an $n \times n$-matrix is created yielding n^2 variables. Hence the more sparse the graph the better performance of the linear relaxation is expected.

| # | node weight | | edge | degree | | density | diameter | algebr. |
nodes	max	mean	weight	max	mean			conn.
100	34	30	14.6	7	2	2.00	14.59	0.01
500	74	50	14.4	10	2	0.40	21.90	0.02
1000	124	75	14.5	11	2	0.20	24.98	0.78
3000	324	175	14.5	12	2	0.07	30.70	1.00
5000	524	275	14.5	13	2	0.04	33.00	1.00

Table 7.14: Characteristic numbers in average of the random trees.

The graphs are generated by the following procedure. The first edge joins some two randomly selected nodes. As long as the number of edges is less than $|V|-1$ we select randomly two nodes and join them by an edge if exactly one of the nodes is isolated. Edge weights vary between 10 and 19 and weights of nodes vary between 25 and $|V|/10$. We generate sets of graphs with $|V| = 100, 500, 3000, 5000$. Each set consists of 100 instances. In Table 7.14 we give the characterization of the graphs in average.

We solve the instances with five different settings depending on the applied formulation of the MGBP and the relaxation: LP(IP2), LP(IP3), LP(IP3)+T, LP(IP3)+SDP, and SDP. The models are described in Section 7.5. Note that, the number of triangle inequalities grows with the increasing density of the graph. In the considered case we obtain less than $|V|$ triangle inequalities. Hence we use such possibility to strengthen the formulation (IP3) of the MGBP.

On the following pages we give the average results for each set of random trees. We compare the average time needed to solve the problems to optimality and the optimality gap in case the optimal solution was not achieved within 4 CPU hours. The numbers are presented in Tables 7.15-7.19. In Figures 7.16-7.20 we depicted the accumulated number of solved instances (the y-axis) within the given time periods (x-axis).

settings	time (sec.)	time geom.	gap (%)
LP(IP2)	0.58	0.39	0
LP(IP3)	1.01	0.57	0
LP(IP3)+T	0.70	0.43	0
LP(IP3)+SDP	5.20	2.25	0
SDP	51.08	8.63	0

Table 7.15: Average time and gap values for random trees with 100 nodes.

All graphs with 100 nodes are solved to optimality with all models except for SDP within 100 seconds. SDP model needs for a few graphs up to half an hour. The shortest amount of time is needed by the model LP(IP2) followed by LP(IP3)+T. All models using exclusively the linear relaxation lead to a short solution time. However, the situation slightly changes with the growing number of nodes in the random trees.

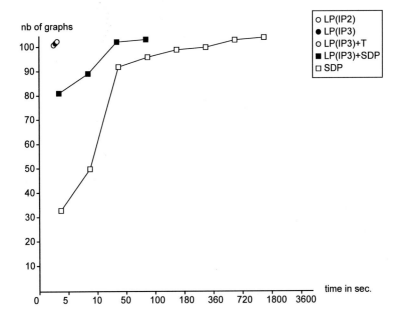

Figure 7.16: Distribution of the computation time for random trees with 100 nodes.

settings	time (sec.)	time geom.	gap (%)	upper bound	lower bound
LP(IP2)	15	9	0	23.91	23.91
LP(IP3)	138	34	0	23.91	23.91
LP(IP3)+T	22	12	0	23.91	23.91
LP(IP3)+SDP	836	73	1	24.01	23.82
SDP	822	287	0	23.91	23.91

Table 7.16: Average values for random trees with 500 nodes.

When we increase the number of nodes up to 500, the pure linear models perform still very well, but one can observe that LP(IP3) is in decline. On average the time for the SDP model to solve random trees with 500 nodes is 16 times longer than the time needed for 100 nodes, while the time of LP(IP3) is over 100 times longer. Hence, the SDP model appears to be less sensitive to the increasing number of nodes in the trees.

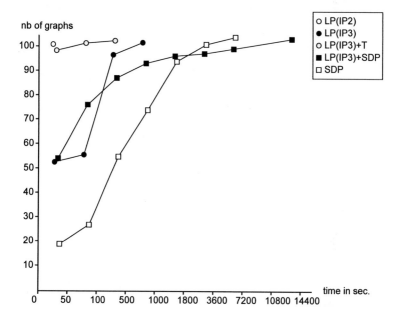

Figure 7.17: Distribution of the computation time for random trees with 500 nodes.

settings	time (sec.)	time geom.	gap (%)	upper bound	lower bound
LP(IP2)	50	28	0	23.46	23.46
LP(IP3)	1328	176	0	23.46	23.46
LP(IP3)+T	75	40	0	23.46	23.46
LP(IP3)+SDP	4312	481	5	24.14	22.90
SDP	4401	1281	2	23.47	23.05

Table 7.17: Average values for random trees with 1000 nodes.

We observe the weakening tendency of the model LP(IP3) with growing number of nodes. For graphs with 1000 nodes the average time to solve the MGBP with LP(IP3) is 20 times longer than with LP(IP2). For graphs with 500 nodes it is only 10 times. Doubling number of nodes causes in average the same growth in the computation time for all models except for LP(IP3), which increases twice as much as the others.

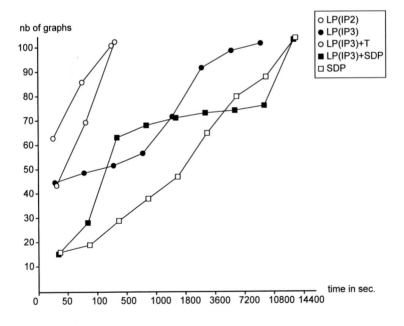

Figure 7.18: Distribution of the computation time for random trees with 1000 nodes.

settings	time (sec.)	time geom.	gap (%)	upper bound	lower bound
LP(IP2)	314	238	0	24.01	24.01
LP(IP3)	9295	2568	26	30.09	24.01
LP(IP3)+T	727	456	0	24.01	24.01
LP(IP3)+SDP	7636	3750	10	25.60	23.21
SDP	11853	8608	21	24.08	20.28

Table 7.18: Average values for random trees with 3000 nodes.

Only the models LP(IP2) and LP(IP3)+T manages to solve all instances with 3000 nodes to optimality within less than given 4 CPU hours, in fact within 2 CPU hours. All other models need more than 4 hours for more than the half of tested graphs. The worst performance is that of LP(IP3) and LP(IP3)+SDP. We observe for the first time that it is worth to combine both models to LP(IP3)+SDP.

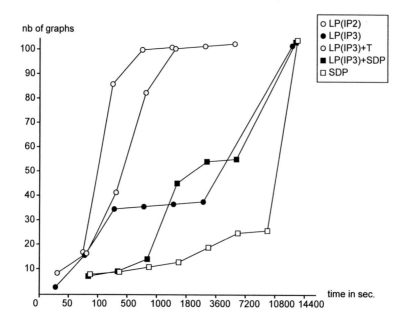

Figure 7.19: Distribution of the computation time for random trees with 3000 nodes.

settings	time (sec.)	time geom.	gap (%)	upper bound	lower bound
LP(IP2)	1036	743	0	24.2	24.2
LP(IP3)	12336	8124	58	36.9	23.8
LP(IP3)+T	2838	1732	1	24.5	24.2
LP(IP3)+SDP	9288	7125	15	27.0	23.3
SDP	12444	10413	43	25.5	18.2

Table 7.19: Average values for random trees with 5000 nodes.

For random trees with 5000 nodes only LP(IP2) solves all instances to optimality. Moreover, the longest computing time takes 2 CPU hours. The presence of the triangle inequalities, as in previous cases, considerably improves the model LP(IP3). On average, it is only twice as slow as LP(IP2) and four times faster than LP(IP3). Furthermore, it manages to solve almost all instances, unlike LP(IP3) with merely 20 units.

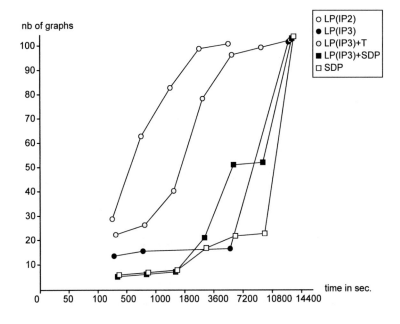

Figure 7.20: Distribution of the computation time for random trees with 5000 nodes.

LP(IP3) and SDP perform almost equally badly with respect to the computing time. Nevertheless, studying Tables 7.18 and 7.19 one observes that the optimality gap for LP(IP3) is due to the high upper bounds, while for the SDP model due to the weakness of the lower bounds. We visualize this behavior in Figures 7.21 and 7.22. For each instance ordered by number we plot the upper and lower bounds normalized with the value of the optimal solution. The dashed line corresponds to the optimal value, which we achieved e.g. by model LP(IP2). As one can see, most of the upper bounds achieved using model LP(IP3), the white filled circles, stay above the dashed line. For only a few instances solved by SDP the upper bounds differ from optimality (gray filled boxes above the dashed line). However, the majority of the gray boxes accumulates below the dashed line, which means that the lower bounds of most instances solved by the SDP model did not reach optimality within 4 CPU hours. The white circles are hardly seen below the dashed line, which means that the LP-relaxation gives almost always the optimal value.

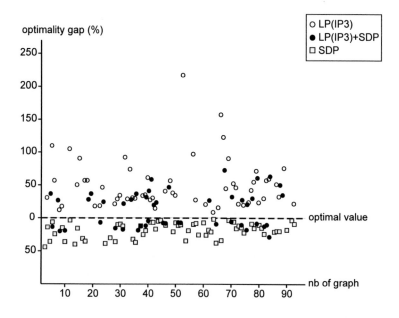

Figure 7.21: Distribution of the optimality gap for random trees with 3000 nodes.

This behavior motivated us to involve the combination of the LP(IP3) and SDP model in form of LP(IP3)+SDP in order to close, or at least to reduce, the optimality gap. As we already mentioned previously, the first calls of the SDP-solver yield good upper bounds. Furthermore, each call of the SDP-solver is time consuming. Thus, in model LP(IP3)+SDP we solve the semidefinite relaxation only several times at the beginning, i.e., in the root node of the b&b tree, to obtain an upper bound of a good quality and then let the LP-relaxation improve the lower bound. The black dots corresponding to the bounds of the LP(IP3)+SDP per instance stay closer to the dashed line. As one can read from Tables 7.18 and 7.19 the reduction of the gap affect obviously the computing time.

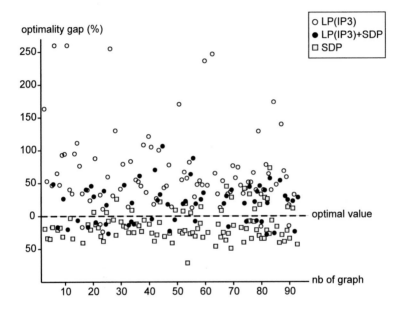

Figure 7.22: Distribution of the optimality gap for random trees with 5000 nodes.

Summing up the tests presented in this section, we observe that very sparse instances can in fact be efficiently solved using the LP-relaxation. However, only the exact formulation as in (IP2) avails. The model (IP3) works well only with the triangle inequalities. But such strengthening we can afford only for very sparse instances. Although the models LP(IP3) and SDP appear to be the losers in the tested sample, we could show that the combination of both the linear and the semidefinite relaxation may be beneficial sometimes.

Chapter 8

Conclusions

In the presented work we consider the minimum graph bisection problem (MGBP), where the nodes of a weighted graph are partitioned into two subsets with prescribed restricted capacity, and the weighted sum of edges joining nodes in different subsets is minimized. This NP-hard combinatorial optimization problem can be modeled by means of linear as well as quadratic integer programming. Each formulation leads to a different kind of approximation, in form of a relaxation, and thus different computational methods applied for their solution. Nevertheless, the feasible sets in each representation are in one-to-one correspondence under an affine transformation. Thus the convex hull of the feasible points, the bisection cut polytope P_B, is a common object to study. The tightest possible description of this polytope can be utilized by solving both relaxations to improve the approximations.

Having presented both mentioned formulations of the MGBP as well as the corresponding linear and semidefinite relaxation in Chapter 2 we attend to polyhedral investigations in the following three chapters.

The first classes of inequalities we look at rely predominantly on the underlying graph structure. Using known approaches of this kind to other graph partitioning problems we exploit the specifics of the MGBP. In particular, we employ the fact that the node set of the graph is partitioned into exactly two subsets. The standard ansatz is to penalize the cases when too many nodes are assigned to one cluster. This is satisfied if, for instance, none of the edges in the subgraph induced by an overweight node set is in the cut. Enforcing that at least one of these edges is in the cut gives rise to a valid inequality for the associated polytope. In the bisection case we can extend the support of such an inequality by identifying edges which taken into a cut lead to an overweight cluster. The corresponding variable enters the inequality in a complementary form. In this way we derive complementarity tree inequalities and complementarity cycles with trees inequalities. Having shown the validity of this inequalities for P_B we design an algorithmic way to determine these inequalities while separating solutions of relaxations to the MGBP.

Considering small graph instances we can identify the complementarity inequalities in the full description of the corresponding bisection cut polytopes. So far, we cannot establish the conditions when these inequalities define facets of P_B. Observing the bad performance of the separation routines during the empirical experiments we assume that one should focus on stronger versions of these inequalities. For this reason we encourage further polyhedral investigations as well as algorithmic improvements.

The next class of valid inequalities for P_B, which we turn to, was introduced by Ferreira et al. in 1996. We take a closer look at these so-called knapsack tree inequalities, as they stroke with computational efficiency by experiments performed by Ferreira et al.. We figure out a strengthening of these inequalities which in special cases leads to the strongest possible form. In particular, we identify necessary and sufficient conditions for the knapsack tree inequalities to define facets of P_B if the underlying graph is a tree with equally weighted nodes.

Although the proof of such a simple case is very complex, we strongly assume that facet-defining property of the knapsack tree inequalities will be preserved if the tree is extended to a denser graph. However, some new conditions have to be figured out concerning the tree supporting the inequality.

The last considered class of inequalities, the bisection knapsack walk inequalities, merges in some sense inequalities presented so far. In particular, the knapsack tree inequalities as well as the complementarity tree inequalities can be seen as special cases of the bisection knapsack walk inequalities. On the one hand, they exploit the knapsack condition on cluster capacities as the knapsack tree inequalities do. On the other hand, they use the bisection case and thus involve variables in the complementary form. The bisection knapsack walk inequalities were introduced by Armbruster in 2007. He also considered a sophisticated strengthening based on an appropriate reduction of the right-hand side. For that he investigated the so-called cluster weight polytope and gave its complete description. We revisit his approach and elaborate an alternative proof for the latter result.

The approximations obtained by relaxations yield lower bounds on the objective function in the integer programming formulations of the MGBP. To obtain upper bounds we apply a heuristic approach. Here, we choose three methods from a variety of non-exact algorithms for solving graph partitioning problems. The first one bases on an evolutionary approach and exploits the solution of a relaxation to construct a feasible solution to the MGBP. We merge this so-called dual information with the primal one in form of the coefficients of objective function in a hybrid version of the algorithm. The second heuristic we develop is a randomized version of the well-known Kernighan-Lin method, called greedy randomized adaptive search procedure. We adjust this procedure to the case when nodes of the graph are non-equally weighted. The last method we take over from an existing software for solving graph partitioning problems arising in scientific computations.

As we already mentioned, we picked up only three algorithms from a long list of heuristics for solving graph partitioning problems. These methods combined with LP-relaxation deliver on average disappointing results in comparison with primal heuristics applied within SDP-solver. As one can see there is still a room for further improvement.

In the tests presented in Chapter 7 we investigate the influence of polyhedral investigations on the tightening of the lower bound of the MGBP. We empirically certify the utility of inequalities, which most luckily define facets of the bisection cut polytope. Furthermore, cuts which use the relationship with the knapsack polytope contribute to the computational efficiency in comparison to cuts depending only on the structure of the underlying graph. However, there is a matter of the interaction with the considered relaxation. The semidefinite relaxation seems to work excellently with the odd cycle cuts, while the linear relaxation benefits from the knapsack tree and the bisection knapsack walk inequalities. Furthermore, we continue the comparison of the semidefinite and the linear relaxation commenced by Armbruster in [4]. The good performance of the semidefinite relaxation remains unperturbed by varying the bisection ratio τ. The linear relaxation seems to benefit from the flexibility of the cluster weights given by growing values of τ. Finally, by means of random instances we showed that combinations of the linear and semidefinite relaxation within one solution process may pay off. To give some numbers, within a few hours we can tackle real world instances up to 240 000 edges. This means, we can solve them to optimality, or at least obtain an approximation of the objective function with an optimality gap less than 2%.

Appendix A

Mathematical Foundations

A.1 Graph Theory

A **graph** $G = (V, E)$ consists of two finite sets, where V is a nonempty set of elements called **nodes** and E is a possibly empty subset of pairs $\{u, v\}$, $u, v \in V$, called **edges**. For short we often write uv for an edge $\{u, v\}$. Two elements u, v of V are **adjacent** if $\{u, v\}$ in E. The edge $\{u, v\}$ is then **incident** on both its **end-nodes** u and v. We say also that the edge $\{u, v\}$ joins or connects nodes u and v. The **degree** of a node $v \in V$, denoted by $\deg(v)$, gives the number of edges incident on the node v. A node $v \in V$ with $\deg(v) = 0$ is called an **isolated** node. If the pairs $\{u, v\} \in E$ are not ordered we say that the graph G is **undirected**. In a **simple** graph any two nodes are connected by at most one edge and none of the nodes is adjacent to itself. A simple and undirected graph is **complete** if $E = \{\{u, v\} : u, v \in V, u \neq v\}$. Every complete graph is also called a **clique**. A complete graph of node cardinality n is usually denoted by K_n.

Given a graph $G = (V, E)$. If a function $f : V \to \mathbb{R}$ or $w : E \to \mathbb{R}$ is defined, i.e., each node $v \in V$ has an associated **node weight** $f(v)$ or each edge $e \in E$ has an associated **edge weight** $w(e)$ then we say that G is a **weighted** graph. For short we write f_v instead of $f(v)$ as well as use the shortcut

$$f(U) := \sum_{v \in U} f_v$$

for some $U \subseteq V$. The same notation holds for the function w and any function defined on a discrete or indexed set.

A **walk** in a graph $G = (V, E)$ is a sequence of vertices and edges $v_0, e_1, v_1, e_2, \ldots, e_n, v_n$, where $v_0, v_1, \ldots, v_n \in V$ and $e_1, \ldots, e_n \in E$, such that vertices and edges adjacent in the sequence are incident. We call v_0 the **start-node**, v_n the **end-node** of the walk and $v_0, v_1, \ldots, v_n$ **inner nodes**. A walk is **closed** if its start-node v_0 equals the end-node v_n. A walk $W = v_0, e_1, v_1, e_2, \ldots, v_n$ is said to **connect** or **join** v_0 and v_n. The number of edges (counting repetitions) n is the **edge length** of the walk, and the number of vertices in the walk $n + 1$ defines the **node length**. The **length** of a walk in an unweighted graph is the edge length and in a weighted graph is the weighted edge length $\sum_{i=1}^{n} w(e_i)$. If no node is repeated in a walk then it is called a **path**. An edge in a path incident on a node of degree 1 is called an **end-edge** of this path. A path in G connecting two nodes u and v with the minimum length is called a **shortest path** from u to v. The **distance** from node u to node v in a graph, denoted $d(u, v)$, is the length of a shortest path from u to v. A **cycle** is a closed path. A cycle having exactly three edges is called a **triangle**.

In a **connected** graph there exists a walk between every pair of vertices.

Given a graph $G = (V, E)$ and sets $U \subseteq V$, $F \subseteq E$. A graph $H = (U, F)$ is called a **subgraph** of G. If $U = V$ then H is a **spanning subgraph** of G. The set

$$E(U) = \{uv \in E : u, v \in U\}$$

is the edge set **induced** by node set U. Then $H = (U, E(U))$ is a **subgraph of G induced by node set** U. Analogously, we say that the set defined as

$$V(F) = \{v \in V : \exists e \in F, e \text{ is incident on } v\}$$

is the node set **induced** by the edges set F, and $H = (V(F), F)$ is a **subgraph of G induced by the edge set** F. A **component** of a graph G is a connected subgraph H such that no subgraph of G that properly contains H is connected.

A **forest** is a graph without cycles. A connected forest is called a **tree**. Given a tree T, an edge in T incident on a node of degree 1 is called a **leaf** and this particular node a **leaf-node**. A **star** is a tree having a node adjacent to all other nodes. This node is called **root** or **center** of the star. A **spanning tree** of a graph G is a spanning subgraph of G that is a tree. A **minimum spanning tree** in G is a spanning tree T with minimal length, i.e., the sum of lengths of all paths in T.
A **shortest path tree** in G from a vertex r is a tree T, **rooted** at r, that contains all the vertices that are reachable from r, in the sense that there is a path in G joining r and some vertex v. The path in T from r to any vertex v is a shortest path in G.
Given a tree $T = (V_T, E_T)$ rooted at some vertex $r \in V_T$. For some edge $e \in E_T$ let Π_e be the path in T joining root r and the edge e. All edges in $e \in E$ having equal number of edges on Π_e, say d, are situated at the same **depth level** in the tree. The corresponding number d is called the **depth** of the tree. The smaller the number d the higher is the depth level. For some path Π in T the **deepest** edge in Π is the leaf of Π non-incident on r.

A graph is **bipartite** if its vertices can be partitioned into two sets in such a way that no edge joins two vertices in the same set. A component H of a graph G is **bipartite connected** if H is bipartite. A node that is isolated in G is considered as a bipartite component in G.

Let $G = (V, E)$ be an undirected graph. A **matching** in G is a set $M \subseteq E$ of pairwise nonadjacent edges. A **perfect matching** is a matching in M in which each vertex of V is incident on exactly one edge of M. In a weighted graph G a **maximum-weight (minimum-weight) matching** of G is a matching M having the largest (smallest) weight $w(M) = \sum_{e \in M} w_e$.

The **adjacency matrix** of a graph $G = (V, E)$ is the $|V| \times |V|$ matrix A with entries

$$A_{ij} = \begin{cases} 1, & ij \in E \\ 0, & \text{otherwise.} \end{cases}$$

The matrix A becomes the **weighted adjacency matrix** of a weighted graph G if one sets $a_{ij} := w_{ij}$ for $ij \in E$. The **Laplace matrix** or shortly **Laplacian** of a graph is defined by

$$L = \text{Diag}(Ae) - A, \qquad e = (1, \ldots, 1)^T. \tag{A.1}$$

If A is the weighted adjacency matrix then L contains the weighted degree of each node on its diagonal and the negative edge weights on its offdiagonal entries. In this case we call the matrix L the **weighted Laplacian**.

For more detailed insight into Graph Theory we refer the interested reader for example to Clark and Holton [18], Gross and Yellen [35].

A.2 Polyhedral Theory

We consider the Euclidean n-dimensional vector space $\mathbb{R}^n$ over the field $\mathbb{R}$ endowed with the inner product

$$x^T y := \sum_{i=1}^{n} x_i y_i$$

for any two vectors $x, y \in \mathbb{R}^n$ and the Euclidean norm

$$||x|| := \sqrt{x^T x}.$$

We denote by e the vector in $\mathbb{R}^n$ with all entries equal to 1 and define $\mathbb{R}^n_+ := \{x \in \mathbb{R}^n : x \geq 0\}$.

A vector $x \in \mathbb{R}^n$ is a **conic combination** of vectors $x_1, \ldots, x_k \in \mathbb{R}^n$ if there exists a vector $\lambda \in \mathbb{R}^k_+$ such that

$$x = \sum_{i=1}^{k} \lambda_i x_i.$$

For a nonempty set $S \subseteq \mathbb{R}^n$ the set of conic combinations of finitely many vectors from S is called **conic hull of S** and denoted by $\text{cone}(S)$. S is called a **cone** if $S = \text{cone}(S)$.

A vector $x \in \mathbb{R}^n$ is an **affine combination** of vectors $x_1, \ldots, x_k \in \mathbb{R}^n$ if there exists a vector $\lambda \in \mathbb{R}^k$ such that

$$\sum_{i=1}^{k} \lambda_i = 1 \quad \text{and} \quad x = \sum_{i=1}^{k} \lambda_i x_i.$$

x is **affinely independent** from vectors $x_1, \ldots, x_k \in \mathbb{R}^n$ if such λ does not exists. The points $x_1, \ldots, x_k \in \mathbb{R}^n$ are affinely independent if and only if the k vectors

$$\begin{pmatrix} x_1 \\ 1 \end{pmatrix}, \ldots, \begin{pmatrix} x_k \\ 1 \end{pmatrix} \in \mathbb{R}^{n+1}$$

are linearly independent. For a nonempty set $S \subseteq \mathbb{R}^n$ the set of affine combinations of finitely many vectors from S is called **affine hull of S** and denoted by $\mathrm{aff}(S)$. S is called an **affine space** if $S = \mathrm{aff}(S)$. S is an **affinely independent set** if each element in S is affinely independent from all other elements in S. The **dimension** of a set $S \subseteq \mathbb{R}^n$, denoted by $\dim(S)$, is the cardinality of a largest affinely independent subset of S.

A set $S \subseteq \mathbb{R}^n$ is called **convex** if for any two vectors $x_1, x_2 \in S$ and any $\lambda \in [0, 1]$ the vector $\lambda x_1 + (1 - \lambda)x_2$ is contained in S. A vector $x \in \mathbb{R}^n$ is a **convex combination** of vectors $x_1, \ldots, x_k \in \mathbb{R}^n$ if there exists a vector $\lambda \in \mathbb{R}^k_+$ such that

$$\sum_{i=1}^{k} \lambda_i = 1 \quad \text{and} \quad x = \sum_{i=1}^{k} \lambda_i x_i.$$

For a nonempty set $S \subseteq \mathbb{R}^n$ the set of all convex combinations of elements in S is called **convex hull of S** and denoted by $\mathrm{conv}(S)$.

For a vector $\alpha \in \mathbb{R}^n \setminus \{0\}$ and a scalar $\alpha_0 \in \mathbb{R}$ the subset $F = \{x \in \mathbb{R}^n : \alpha^T x = \alpha_0\}$ is called **hyperplane** and the subset $H = \{x \in \mathbb{R}^n : \alpha^T x \le \alpha_0\}$ is called **halfspace**. F and H are **induced** by the inequality $\alpha^T x \le \alpha_0$.

A convex set $P \subseteq \mathbb{R}^n$ is a **polyhedron** if it is finitely generated, that is, there are finite sets $V, E \in \mathbb{R}^n$ such that
$$P = \mathrm{conv}(V) + \mathrm{cone}(E).$$

Equivalently, a polyhedron can be described as an intersection of finitely many halfspaces, i.e., there is a matrix $A \in \mathbb{R}^{m \times n}$ and a vector $b \in \mathbb{R}^m$ such that

$$P := P(A, b) = \{x \in \mathbb{R}^n : Ax \le b\}.$$

This is the assertion of the well known **representation theorem**, see for instance Grötschel [40] and Ziegler [89]. If a polyhedron P is bounded, i.e., there is a $c \in \mathbb{R}_+ \setminus \{0\}$ such that $P \subseteq \{x \in \mathbb{R}^n : ||x|| \le c\}$, we call it **polytope**. A cone is **polyhedral** if it is a polyhedron. The **dimension** of a polyhedron P, denoted by $\dim(P)$, is one less than the maximum number of affinely independent vectors in P. A polyhedron $P \subseteq \mathbb{R}^n$ is **full-dimensional** if $\dim(P) = n$.

Given a vector $\alpha \in \mathbb{R}^n \setminus \{0\}$ and a scalar $\alpha_0 \in \mathbb{R}$, an inequality $\alpha^T x \le \alpha_0$ is **valid** for a polyhedron P if the halfspace induced by this inequality includes P. A set $F_\alpha \subseteq P$ defined by the intersection of P and the hyperplane induced by a valid inequality $\alpha^T x \le \alpha_0$ for P,

$$F_\alpha = P \cap \{x \in \mathbb{R}^n : \alpha^T x = \alpha_0\},$$

is called a **face** of P. The inequality $\alpha^T x \le \alpha_0$ **represents** or **defines** the face F. F is a **proper** face of P if $F \ne P$ and **nontrivial** if additionally $\emptyset \ne F$. A nontrivial face F of a polyhedron P which is not included in any other proper face of P is called a **facet**. An equivalent definition reads: A face F of P is called a **facet** if $\dim(F) = \dim(P) - 1$. A face of dimension 0 is called an **extreme point** in P.

Let $\alpha^T x \le \alpha_0$ and $\beta^T x \le \beta_0$ be valid inequalities for a polytope $P \subset \mathbb{R}^n$. We say that the inequalities are **equivalent** if there exists a scalar $\rho \in \mathbb{R} \setminus \{0\}$ such that $\alpha = \rho\beta$ and $\alpha_0 = \rho\beta_0$. Otherwise they are **not equivalent**. If $\alpha^T x \le \alpha_0$ and $\beta^T x \le \beta_0$ are not equivalent and there exists $\rho > 0$, $\rho \in \mathbb{R}$, such that $\alpha \ge \rho\beta$ and $\alpha_0 \le \rho\beta_0$ then we say that inequality $\alpha^T x \le \alpha_0$ **dominates** or **is stronger than** inequality $\beta^T x \le \beta_0$. On the other side $\beta^T x \le \beta_0$ is **dominated by** or **weaker than** $\alpha^T x \le \alpha_0$. Note that if $\alpha^T x \le \alpha_0$ dominates $\beta^T x \le \beta_0$ then

$$\{x \in \mathbb{R}^n_+ : \alpha^T x \le \alpha_0\} \subseteq \{x \in \mathbb{R}^n_+ : \beta^T x \le \beta_0\}.$$

A procedure to obtain a stronger inequality than a given one is called a **strengthening**, see e.g. Proposition A.7 below. As consequences from the above definitions one obtains the following propositions.

Proposition A.1 *Let $\alpha^T x \le \alpha_0$ be a valid inequality for a polyhedron $P(A, b)$ inducing the face $F_\alpha = \{x \in P : \alpha^T x = \alpha_0\}$. Then F_α is a facet of P if and only if for any other valid inequality $\beta^T y \le \beta_0$ such that $F_\alpha \subseteq \{x \in P : \beta^T x = \beta_0\}$ there exists $\gamma \ge 0$ and a vector u (of appropriate dimension) such that*

$$\begin{pmatrix} \beta \\ \beta_0 \end{pmatrix} = \gamma \begin{pmatrix} \alpha \\ \alpha_0 \end{pmatrix} + \begin{pmatrix} A^T \\ b^T \end{pmatrix} u.$$

In particular, if P is full-dimensional, i.e., aff$(P) = \{0\}$, then F_α is a facet if and only if for any other valid inequality $\beta^T x \le \beta_0$ with $F_\alpha \subseteq \{x \in P : \beta^T x = \beta_0\}$ there exists $\gamma \ge 0$ such that

$$\begin{aligned} \beta &= \gamma\alpha, \\ \beta_0 &= \gamma\alpha_0. \end{aligned}$$

The second part of the above proposition can be generalized as follows.

Proposition A.2 *Let $F_\alpha = \{x \in P : \alpha^T x = \alpha_0\}$ be a proper face of a full-dimensional polyhedron P. The following statements are equivalent:*

(1) F_α is a facet of P.

(2) If $\beta^T x = \beta_0$ for all $x \in F$ then for some $\gamma \geq 0$ it holds

$$\begin{aligned} \beta &= \gamma\alpha, \\ \beta_0 &= \gamma\alpha_0. \end{aligned}$$

For the proofs of the above two propositions we refer to Grötschel [40] or Nemhauser and Wolsey [72]. We state below some technical results following from the last proposition.

Corollary A.3 *Let $\alpha^T x \leq \alpha_0$ and $\beta^T x \leq \beta_0$ be valid and non-equivalent inequalities for a full-dimensional polyhedron $P \subset \mathbb{R}^n$. If $\beta x = \beta_0$ holds for all $x \in F_\alpha$ then F_α does not define a facet of P.*

Corollary A.4 *Let $\alpha^T x \leq \alpha_0$ and $\beta^T x \leq \beta_0$ be valid inequalities for a full-dimensional polyhedron $P \subset \mathbb{R}^n$. If the inequalities are not equivalent then the face of P defined by the inequality $(\alpha + \beta)^T x \leq \alpha_0 + \beta_0$ is not a facet.*

Proof. It is obvious that $(\alpha + \beta)^T x \leq \alpha_0 + \beta_0$ is valid for P. Let F_α, F_β and $F_{\alpha+\beta}$ be faces of P defined by the inequalities $\alpha^T x \leq \alpha_0$, $\beta^T \leq \beta_0$ and $(\alpha + \beta)^T x \leq \alpha_0 + \beta_0$, respectively. Any $x \in P$ such that $x \in F_\alpha$ and $x \notin F_\beta$ does not lie in $F_{\alpha+\beta}$ as well. The same holds for $x \in P$ such that $x \notin F_\alpha$ and $x \in F_\beta$. This yields $F_{\alpha+\beta} = F_\alpha \cap F_\beta$. Therefore $\alpha^T \overline{x} = \alpha_0$ holds for all $\overline{x} \in F_{\alpha+\beta}$ and hence $F_{\alpha+\beta}$ cannot define a facet of P by Corollary A.3. $\square$

Proposition A.5 *Let $\sum_{i=1}^n \alpha_i x_i \leq \alpha_0$ with $\alpha_{k+1} = \alpha_{k+2} = \ldots = \alpha_n = 0$ for some $k < n$ be a facet defining inequality for the polytope $P \subset \mathbb{R}^n$. Let P' be a projection of P onto the space $\mathbb{R}^k$ such that $P' := \{(x_1, \ldots, x_k)^T : (x_1, \ldots, x_n)^T \in P\}$. Then*

$$\sum_{i=1}^k \alpha_i x_i \leq \alpha_0$$

defines a facet of P'.

For the proof see e.g. Grötschel [40].

We introduce now some non-standard but useful abbreviations concerning valid inequalities for polyhedrons in general. Let $\alpha^T x \geq \alpha_0$ be a valid inequality for a polyhedron $P \subset \mathbb{R}^n$ and $\overline{x} \in P$. If $\alpha^T \overline{x} = \alpha_0$ we say that the vector $\overline{x}$ is **tight** for the inequality or that $\overline{x}$ is a **root** of the inequality. The set of indices I such that $\alpha_i \neq 0$ for each $i \in I$ is called **support of the inequality** $\alpha^T x \geq \alpha_0$.

Finally we cite the so-called **separation theorem**, which legitimates the cutting planes approach in solving integer programs, see Section A.3. For the proof of this theorem see e.g. Grötschel [40].

Theorem A.6 (separation theorem for polyhedra) *Let* $P(A, a)$ *and* $Q(B, b)$ *be polyhedra in* $\mathbb{R}^n$ *such that* $P \cap Q = \emptyset$ *and* $P \neq \mathbb{R}^n \neq Q$. *Then there exists a vector* $\gamma \in \mathbb{R}^n \setminus \{0\}$ *and a scalar* $\gamma_0 \in \mathbb{R}$ *such that*

$$P \subseteq \{x \in \mathbb{R}^n : \gamma^T x < \gamma_0\} \quad and \quad Q \subseteq \{x \in \mathbb{R}^n : \gamma^T x > \gamma_0\},$$

i.e., the hyperplane $H = \{x \in \mathbb{R}^n : \gamma^T x = \gamma_0\}$ *separates strongly* P *and* Q.

0-1 Polytopes

Given a finite set $S \subseteq \{0, 1\}^n$ we call $P = \text{conv}(S)$ a **binary** or a **0-1 polytope**. The following proposition delivers a tool for the strengthening of valid inequalities for binary polytopes.

Proposition A.7 *Let* $S \subseteq \{0, 1\}^n$ *be finite,* $P = \text{conv}(S)$ *and* $\alpha^T x \geq \alpha_0$ *be a valid inequality for* P. *Define*

$$\beta_i := \min\{\alpha_i, \max\{0, \alpha_0 - \sum_{j : \alpha_j < 0} \alpha_j\}\}, \qquad i = 1, \dots, n. \tag{A.2}$$

Then the inequality $\beta^T x \geq \alpha_0$ *dominates* $\alpha^T x \geq \alpha_0$ *and is valid for* P.

Proof. Let $\overline{x} \in P \cap \{0, 1\}^n$. If $\overline{x}_i = 0$ for all i such that $\beta_i \neq \alpha_i$ then trivially $\beta^T \overline{x} \geq \alpha_0$. Otherwise $\overline{x}_j = 1$ for some j, $1 \leq j \leq n$, with $\beta_j \neq \alpha_j$. Then

$$\sum_{i=1}^n \beta_i \overline{x}_i - \alpha_0 = \sum_{i : \beta_i \geq 0} \beta_i \overline{x}_i + \sum_{i : \beta_i < 0} \beta_i \overline{x}_i - \alpha_0 \geq \beta_j - \left(\alpha_0 - \sum_{i : \alpha_i < 0} \alpha_i\right)$$

$$= \max\{0, \alpha_0 - \sum_{i : \alpha_i < 0} \alpha_i\} - \left(\alpha_0 - \sum_{i : \alpha_i < 0} \alpha_i\right) \geq 0,$$

where the first inequality follows from the fact that $\beta_i = \alpha_i$ if $\alpha_i < 0$. $\qquad \square$

For further details on Polyhedral Theory we refer the reader to Grötschel [40], Nemhauser and Wolsey [72], or Ziegler [89].

A.3   Integer Programming

We assume that the reader is familiar with linear programming, otherwise we refer to Vanderbei [84] for a profound introduction. Suppose we have a **linear program**

$$(\text{LP}) \qquad\qquad \min\{c^T x \: : \: Ax \leq b, \: x \in \mathbb{R}^n_+\},$$

where A is an $m \times n$ matrix with entries in $\mathbb{R}$, $c \in \mathbb{R}^n$ and $b \in \mathbb{R}^m$, and $x \in \mathbb{R}^n$ is the vector of **variables** or **unknowns**. If all variables are integers we have a **linear integer program**,

$$(\text{ILP}) \qquad\qquad \min\{c^T x \: : \: Ax \leq b, \: x \in \mathbb{Z}^n_+\},$$

and if all variables are restricted to 0-1 values we have a **linear 0-1** or **binary program**

$$(\text{BLP}) \qquad\qquad \min\{c^T x \: : \: Ax \leq b, \: x \in \{0,1\}^n\}.$$

Similarly, suppose we have a **quadratic program**

$$(\text{QP}) \qquad\qquad \min\{\tfrac{1}{2}x^T Q x + c^T x \: : \: Ax = b, \: x \in \mathbb{R}^n\},$$

where Q is a symmetric $n \times n$ matrix with entries in $\mathbb{R}$, $A \in \mathbb{R}^{m \times n}$, $c \in \mathbb{R}^n$ and $b \in \mathbb{R}^m$ and $x \in \mathbb{R}^n$ is a vector of variables or unknowns. If all variables are integers we have a **quadratic integer program**,

$$(\text{IQP}) \qquad\qquad \min\{\tfrac{1}{2}x^T Q x + c^T x \: : \: Ax = b, \: x \in \mathbb{Z}^n\}.$$

Although linear and quadratic problems can be solved in polynomial time, provided the matrix Q is positive semidefinite, restricting the variables to discrete domains yields NP-hard problems in general. Most combinatorial optimization problems can be stated as binary linear or integer quadratic programs. Consider an **integer program** in a general form

$$(\text{IP}) \qquad\qquad z := \min\{\, c(x) \: : \: x \in Z \subseteq \mathbb{Z}^n\},$$

where $c(x)$ is either a linear or quadratic function and Z is the set of feasible solutions defined as an intersection of finitely many halfspaces and $\mathbb{Z}^n$. A "naive" but nonetheless important approach to solve (IP) is to find a lower bound $\underline{z} \leq z$ and an upper bound $\overline{z} \geq z$ such that $\underline{z} = \overline{z} = z$. Every feasible solution $x^* \in Z$ provides an upper bound. Finding upper bounds is generally tackled heuristically. A common approach to obtain a lower bound is to **relax** the problem, i.e., to replace a difficult (IP) by a simpler optimization problem. There are two obvious possibilities to obtain a **relaxation**. Either one enlarges the set of feasible solutions or one replaces the objective function by a function that has a smaller value for all feasible solutions. It is obvious that the optimal value of a relaxation of (IP) is always less or equal than z. A straight forward relaxation for a linear integer program (ILP) is the **linear programming relaxation** (LP), where the integrality of the variables is relaxed to real values. Similarly, a linear programming relaxation to a (BLP) reads

(LBP) $$\min\left\{c^T x \ : \ Ax \leq b, \ x \in [0,1]^n\right\}.$$

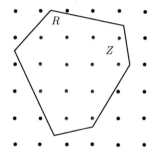 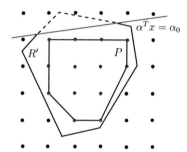

Figure A.1: The set of feasible integer solutions Z (the interior dots) and the feasible set of the relaxation R.

Figure A.2: A valid inequality $\alpha^T x \geq \alpha_0$ for the integral polytope P tightening the relaxation R.

In often turns out that the basic relaxation delivers a weak approximation of the objective function z. An obvious way to improve the lower bound is to narrow the feasible set of the relaxation. This is usually achieved by deriving valid inequalities for the **integer polyhedron** defined by $P := \mathrm{conv}(Z)$, see Figure A.1 and A.2. In this context one considers a related problem to (IP), called the separation problem.

Problem A.8 (separation problem) *Given a solution of a relaxation to (IP), say $\bar{x}$, is $\bar{x} \in \mathrm{conv}(Z)$? If not, find an inequality $\alpha x \leq \alpha_0$ satisfied by all points in Z, but violated by the point $\bar{x}$.*

Remark A.9 *Grötschel, Lovász and Schrijver [41] showed in 1988 the equivalence between the optimization and the separation problem. The result applies for optimization problems with a linear objective function and linear contraints.*

The linearization of the quadratic objective function in (IQP) by considering the matrix $X = xx^T$ leads to a semidefinite programming relaxation (see Section A.4 below). As we showed in Section 2.5 this relaxation can be also sharpened by valid inequalities for the associated integer polyhedron.

A natural starting point in solving linear integer programs (ILP) with integral data (A, b) is to ask when the linear programming relaxation (LP) has an optimal solution that is integral. One obtains a positive answer if the matrix A is specially structured.

A matrix A is **totally unimodular** if every square submatrix of A has determinant $+1$, -1 or 0.

Proposition A.10 *For a linear integer program (ILP) let $Z := \{x \in \mathbb{Z}_+^n : Ax \leq b\}$ and $P := conv(Z)$. If b is an integer vector and A is totally unimodular then all vertices of the polyhedron P are integral, i.e., P is fully described by the inequalities $Ax \leq b$.*

For the proof see Wolsey [88].

Branch-and-Cut Algorithm

Branch-and-cut algorithms follow a divide-and-conquer strategy and are currently among the most successful methods to solve integer programs to optimality. Branch-and-cut algorithm originates from the **branch-and-bound** method. It is basically an enumeration approach in a fashion that it identifies and prunes non-promising parts of search space. The method was introduced by Land and Doig in 1960 for integer linear programming. The two main components of the method are **branching**, dividing the feasible region into subregions, and **bounding** which concerns finding upper and lower bounds for the optimal solution within a feasible subregion. All produced subregions can be naturally stored in a tree structure called **branch-and-bound tree**. Its nodes correspond to the constructed subregions. The success of the method relies on the quality of obtained bounds. A standard technique to improve lower bounds in linear integer programming is based on strengthening the initial linear relaxation by adding additional constraints satisfied by all integer solutions. Such a linear constraint corresponds to a hyperplane called a **cutting plane**, for reasons we explain below. **Cutting plane algorithms** are designed to find such constraints. This approach is taken over by other types of relaxations, e.g., a semidefinite relaxation. A **branch-and-cut algorithm** is a branch-and-bound algorithm equipped with a cutting plane algorithm. Applied for solving a binary linear program (BLP) it works as follows, see also Algoriithm A.1.

Denote by X the set of feasible solutions to (BLP),

$$X := conv\{x \in \{0,1\}^n : Ax \leq b\} \cap \{0,1\}^n .$$

Algorithm A.1 Branch-and-Cut

Input A linear binary problem (BLP) Π_0.

1: Initialize the upper bound for Π_0, $u_b := +\infty$; $\quad \Pi := \Pi_0$;

2: **repeat**

3:     Solve the relaxation of Π ; Let x_Π be an optimal solution ;

4:     **if** x_Π is not integral **then**

5:         Look for a violated inequality that cuts off x_Π ;

6:         If a cutting plane is found, update Π ;

7:     **end if**

8: **until** x_Π is integral **or** no cutting plane found

9: **if** x_Π is not integral **then**

10:     Split Π into subproblems and add them to the list of subproblems $\mathcal{S}$;

11:     Update the lower bound, $l_b := \min\{c^T x_\Pi : \Pi \in \mathcal{S}\}$;

12:     **while** $\mathcal{S}$ not empty **and** $l_b < u_b$ **do**

13:         Choose and remove problem Π from $\mathcal{S}$;

14:         **if** $c^T x_\Pi \geq u_b$ **then**

15:             **go to** Step 12 ;

16:         **else**

17:             Perform the cutting plane phase for Π: Step 2 - 8 ;

18:             Perform the dividing phase for Π: Step 9 - 11 ;

19:             **if** x_Π is a feasible integral solution **and** $c^T x_\Pi < u_b$ **then**

20:                 $u_b := c^T x_\Pi$; $\quad x_{OPT} := x_\Pi$;

21:             **end if**

22:         **end if**

23:     **end while**

24: **else**

25:     $u_b := c^T x_\Pi$; $\quad x_{OPT} := x_\Pi$;

26: **end if**

Output x_{OPT}, an optimal solution to Π_0 or message **problem infeasible** if $u_b = +\infty$.

We assume that $X \neq \emptyset$ holds for our problem (BLP). The first step of the algorithm is to consider a relaxation of (BLP) by choosing a set $R \subseteq \mathbb{R}^n$ with $X \subseteq R$ and to optimize the linear objective function over R. For example, this relaxation might be the linear relaxation (LBP) or a semidefinite programming relaxation (see Section A.4). Let $\overline{x}$ be an optimal solution to the relaxation R. If $\overline{x} \in \{0,1\}^n$ and the set of solutions to the system $Ax \leq b$ is fully included in R, we have an optimal solution for Π. Otherwise, there exists a hyperplane $\{x \in \mathbb{R}^n : \alpha^T x = \alpha_0\}$ such that $\alpha^T \bar{x} > \alpha_0$ and $\mathrm{conv}(X) \subseteq \{x \in \mathbb{R}^n : \alpha^T x \leq \alpha_0\}$ (as it follows from Theorem A.6). Such a hyperplane is called a **cutting plane**. We say that the valid inequality $\alpha^T x \leq \alpha_0$ is **violated** by $\bar{x}$. If we are able to find such a cutting plane, that is, we are able to solve the so-called **separation problem**, we can strengthen the relaxation and continue. This process is iterated until $\bar{x} \in \{0, 1\}^n$ or no more cutting planes are found. If this so-called **cutting plane phase** finishes without finding an optimal solution to Π the **enumeration** or **dividing phase** begins. Some variable with a fractional value $\bar{x}_i$ is selected and two subproblems are created, where one additionally requires $x_i = 0$ and the other $x_i = 1$.

The subproblems are handled in the same manner as the initial problem Π. In the worst case we are forced to branch at each variable x_i and thus solve 2^n subproblems. The list of subproblems can be shortened by maintaining the lower and upper bound of the initial problem Π. Each integral solution $\bar{x}$ to a subproblem delivers an upper bound for the objective function value of Π. The solution with the smallest objective value gives the best upper bound u_b. Suppose that an optimal solution x' of the relaxation of a subproblem Π' satisfies $c^T x' \geq u_b$. Then subproblem Π' and further subproblems derived from Π' do not need to be considered because the optimal solution of these subproblems cannot be better than the best 0-1 solution corresponding to u_b. On the other hand, the value

$$l_b = \min\{c^T \bar{x} : \bar{x} \text{ is an optimal solution to a subproblem of } \Pi\}$$

gives a lower bound for the objective function value of Π. If at some point we obtain $u_b = l_b$ then we have a proof that the solution corresponding to u_b is the optimal solution to Π, because considering further open subproblems may only result in integral solutions with higher objective values than u_b.

In the cutting plane phase usually several **separation algorithms** are called. They are designed with respect to the special structure of the corresponding class of valid inequalities. A separation algorithm for a given class of inequalities solves the following problem.

Problem A.11 (separation of an inequality class) *Given a solution of a relaxation $\bar{x}$ and a class $\mathcal{I}$ of inequalities. Either show that $\bar{x}$ satisfies all inequalities in $\mathcal{I}$ or find an inequality in $\mathcal{I}$ violated by $\bar{x}$.*

As already mentioned, the aim of the cutting plane phase is to strengthen the relaxation and thus improve the lower bound. Generally, if the relaxation to (BLP) is of a poor quality then a large amount of subproblems must be solved till a feasible integral solution, and thus the upper bound, is found. To establish or improve the upper bound one can apply heuristic methods to find some feasible 0-1 solution either by making use of the structure of the underlying problem or by a subtle rounding of relaxation's solutions. At which step of the branch-and-cut algorithm a heuristic is called depends on the information it uses to produce a solution. Since such methods usually act on the initial (primal) formulation of the problem they are called **primal heuristics** or **primal methods**.

We refer the reader to Nemhauser and Wolsey [72, 88] for full details on linear integer programing and to Nocedal and Wright [74] for quadratic programming.

A.4 Semidefinite Programming

In the sequel we give a brief introduction to semidefinite programming and explain the spectral bundle method, an algorithm applied for solving semidefinite programs. We strongly rely on Helmberg [42, 43, 45, 44], where the topics are expatiated.

A.4.1 Positive Semidefinite Matrices

The set $M_{m,n}$ of $m \times n$ real matrices can be interpreted as a vector space in $\mathbb{R}^{n \cdot m}$. In this space the inner product between two elements $A, B \in M_{m,n}$ is defined by

$$\langle A, B \rangle := \mathrm{vec}(B)^T \mathrm{vec}(A) = \sum_{i=1}^{m} \sum_{j=1}^{n} a_{ij} b_{ij} = \mathrm{tr}(B^T A),$$

where the **trace** $\mathrm{tr}(\cdot)$ is the sum of the diagonal elements of a square matrix and the vec-operator $\mathrm{vec}(\cdot)$ is defined via

$$\mathrm{vec}(A) = \begin{bmatrix} A_{.,1} \\ \vdots \\ A_{.,n} \end{bmatrix}$$

and called the **vector representation of** A. We denote by M_n the subset of square matrices in $M_{m,n}$. A symmetric matrix $A \in S_n$ is **positive semidefinite** ($A \in S_n^+$ or $A \succeq 0$) if $x^T A x \geq 0$ for any $x \in \mathbb{R}^n$. A is **positive definite** ($A \in S_n^{++}$ or $A \succ 0$) if $x^T A x > 0$ for any $x \in \mathbb{R}^n \setminus \{0\}$. Next, we list a few useful characterizations of positive (semi)definite matrices.

Proposition A.12 *Let $B \in M_n$ be a nonsingular matrix (i.e., B^{-1} exists). Then $A \in S_n^+$ if and only if $B^T A B \in S_n^+$ and $A \in S_n^{++}$ if and only if $B^T A B \in S_n^{++}$.*

Theorem A.13 (characterizations of positive semidefinite matrices) *For $A \in S_n^+$ the following statements are equivalent:*

(i) A is positive semidefinite.

(ii) All eigenvalues of A are nonnegative, i.e., $\lambda_i(A) \geq 0$, $i = 1, \ldots, n$.

(iii) There exists $C \in M_{m,n}$ so that $A = C^T C$. For any such C, $\mathrm{rank}(C) = \mathrm{rank}(A)$.

(iv) $\langle A, B \rangle \geq 0$ for all $B \in S_n^+$.

The equivalence $(i) \Leftrightarrow (iv)$ is also known as **Frejer's trace theorem**. The set of positive semidefinite matrices is a full-dimensional cone in $\mathbb{R}^{\binom{n+1}{2}}$. A convex set $F \subseteq C$ is called a **face** of a convex set C if for any two elements $x, y \in C$ with $\alpha x + (1 - \alpha) y \in F$ for some $\alpha \in (0, 1)$ we have $x, y \in F$. The following theorem gives a characterization of the faces of the semidefinite cone.

Theorem A.14 (face characterization of the semidefinite cone) *A set $F \neq \emptyset$ is a face of S_n^+ if and only if $F = \{0_{n \times n}\}$ or $F = \{X \ : \ X = P W P^T, W \in S_k^+\}$ for some $k \in \{1, \ldots, n\}$, $P \in M_{n,k}$ with $\mathrm{rank}(P) = k$.*

A.4.2   Semidefinite Programs

Semidefinite programming is linear programming over the cone of semidefinite matrices. In comparison to standard linear programming the cone of the nonnegative orthant $\mathbb{R}_+^n$ is replaced by the cone of semidefinite matrices S_n^+. Semidefinite programs arise in a natural way from problems whose data are given by matrices. The standard formulation of a **semidefinite program** reads

$$\min \langle C, X \rangle,$$

$$\text{s.t.} \quad \langle A_1, X \rangle = b_1,$$

$$\vdots$$

$$\langle A_m, X \rangle = b_m,$$

$$X \succeq 0,$$

where $A_i \in S_n$, $i = 1, \ldots, m$. For ease of exposition one collects the constraints in a linear operator $\mathcal{A} : S_n \to \mathbb{R}^m$,

$$\mathcal{A}(X) = \begin{pmatrix} \langle A_1, X \rangle \\ \vdots \\ \langle A_m, X \rangle \end{pmatrix},$$

and simply writes

(SDP) $\qquad\qquad\qquad \min \{\langle C, X \rangle \ : \ \mathcal{A}(X) = b, X \succeq 0\}.$

To expose the connections with linear programming one can rewrite (SDP) in an equivalent "linear" form

(LSDP) $\min \{ c^T \text{vec}(X) \ : \ A \, \text{vec}(X) \leq b, \ X \succeq 0 \}$,

where $A \in M_{m \times n^2}$ and row $A_{i,\cdot}$ corresponds to the symmetric matrix A_i in (SDP), i.e., $\langle A_i, X \rangle := A_{i,\cdot} \cdot \text{vec}(X)$.

Similar to linear programming the duality theory plays an important role in semidefinite optimization. In order to derive the dual problem to (SDP) we need the adjoint operator to $\mathcal{A}$, which we denote by $\mathcal{A}^T : \mathbb{R}^m \to S_n$. Since

$$\langle \mathcal{A}(X), y \rangle = \sum_{i=1}^m y_i \langle A_i, X \rangle = \langle \sum_{i=1}^m y_i A_i, X \rangle$$

for all $X \in S_n$ and $y \in \mathbb{R}^m$, it has the form

$$\mathcal{A}^T(y) = \sum_{i=1}^m y_i A_i. \tag{A.3}$$

The dual program to (SDP) is obtained via a Lagrange approach by interchanging inf and sup

$$\inf_{X \succeq 0} \sup_{y \in \mathbb{R}^m} \langle C, X \rangle + \langle b - \mathcal{A}(X), y \rangle \geq \sup_{y \in \mathbb{R}^m} \inf_{X \succeq 0} \langle b, y \rangle + \langle X, C - \mathcal{A}^T(y) \rangle. \tag{A.4}$$

The supremum on the right hand side remains finite if the inner minimization over $X \succeq 0$ is finite for some $y \in \mathbb{R}^m$. By Theorem A.13 this is satisfied if $C - \mathcal{A}^T(y)$ is positive semidefinite. We write this condition using a slack matrix Z

(DSDP)
$$\begin{aligned} \max \ &\langle b, y \rangle, \\ \text{s.t.} \quad &\mathcal{A}^T(y) + Z = C, \\ &y \in \mathbb{R}^m, \\ &Z \succeq 0. \end{aligned}$$

This is a standard formulation of the **dual semidefinite program** to (SDP). The gap between a dual feasible solution (y, Z) and a primal feasible solution X is

$$\langle C, X \rangle - \langle b, y \rangle = \langle \mathcal{A}^T(y) + Z, X \rangle - \langle \mathcal{A}(X), y \rangle = \langle Z, X \rangle \geq 0$$

by (A.4) and gives us the **weak duality**. If $\langle Z, X \rangle$ turns out to be zero then this primal-dual pair is an optimal solution. However, in contrast to linear programming it is no longer true that optimality implies $\langle Z, X \rangle = 0$. This behavior of the primal-dual pair is due to the pure algebraic dualization procedure which does not take into account the geometry of the feasible sets. The gap between the optimal primal and the dual objective value is guaranteed to be zero (**strong duality**) if either (SDP) or (DSPD) or both have **strictly feasible** solutions, i.e., (SDP) has a feasible solution $X \in S_n^{++}$ and (DSDP) has a feasible solution (y, Z) such that $Z \in S_n^{++}$, respectively.

The assumption of the existence of a strictly feasible point is a Slater-type regularity condition, a sufficient condition for strong duality in general convex programming, see e.g. Rockafellar [78].

A.4.3   Spectral Bundle Method

By Theorem A.13 a symmetric matrix $X \in S_n$ is positive semidefinite if and only if the minimal eigenvalue of X is nonnegative. Due to this property semidefinite programming and eigenvalue optimization over affine sets of matrices are tightly related.

Proposition A.15 (eigenvalue representation of semidefinite programs)
Consider the primal-dual pair of semidefinite programs

$$\text{(P)} \quad \begin{array}{ll} \max & \langle C, X \rangle, \\ \text{s.t.} & \mathcal{A}(X) = b, \\ & X \succeq 0, \end{array} \qquad \text{(D)} \quad \begin{array}{ll} \min & \langle b, y \rangle, \\ \text{s.t.} & Z = \mathcal{A}^T(y) - C, \\ & y \in \mathbb{R}^m, \ Z \succeq 0. \end{array}$$

Assume there exists $\overline{y} \in \mathbb{R}^m$ with $I = \mathcal{A}^T(\overline{y})$. Then (D) is equivalent to

$$\min_{y \in \mathbb{R}^m} a\lambda_{\max}(C - \mathcal{A}^T(y)) + b^T y \qquad (A.5)$$

for $a = \max\{0, b^T \overline{y}\}$. Furthermore, if (SDP) is feasible then all feasible solutions X satisfy $\operatorname{tr}(X) = a$, the primal optimum is attained and it is equal to the infimum of (DSDP).

For the proof of this proposition we refer to Helmberg [42]. We just outline the idea. The assumption allows to write the constraint $Z \succeq 0$ in (DSDP) as $0 \geq -\lambda_{\min}(Z) = \lambda_{\max}(-Z)$ and to lift it into the objective function by means of a Lagrange multiplier $a \geq 0$.

Spectral Bundle Method with Bounds

In our empirical studies we solve the semidefinite relaxation to the MGBP using the spectral bundle method with bounds equipped with a cutting plane algorithm. We briefly introduce this algorithm below. For an overview on other methods for solving semidefinite programming problems we refer to Armbruster [4].

Due to Proposition A.15 a large class of semidefinite programs can be equivalently formulated as **eigenvalue optimization problems** of the form

(E) $$\min_{y \in \mathbb{R}^m} \lambda_{\max}(C - \mathcal{A}^T(y)) + b^T y.$$

This concerns several important relaxations of combinatorial optimization problems, in particular, the minimum graph bisection problem. Since

$$\lambda_{\max}(A) = \max_{v^T v = 1} v^T A v = \max_{v^T v = 1} \langle A, v^T v \rangle$$

and

$$\text{conv}\{vv^T \, : \, v^t v = \text{tr}(vv^T) = 1\} = \{W \succeq 0 \, : \, \text{tr} W = 1\} =: \mathcal{W},$$

the maximum eigenvalue function may equivalently be formulated as a semidefinite program

$$\lambda_{\max}(A) = \max\{\langle A, W \rangle \, : \, W \in \mathcal{W}\}. \tag{A.6}$$

This characterization of $\lambda_{\max}(\cdot)$ as the maximum over a family of linear functions implies that $\lambda_{\max}(\cdot)$ is a nonsmooth convex function. Therefore techniques from nonsmooth convex optimization can be employed to solve (E). The spectral bundle method is a specialized subgradient method of Kiwiel [59]. In the smooth nonlinear optimization the gradient plays a dominant role. The corresponding object in the nonsmooth convex optimization is the subgradient of a convex function f. Let $D \subseteq \mathbb{R}^n$ and $f \, : \, D \to \mathbb{R} \cup \{+\infty, -\infty\}$ be a convex function. A **subgradient** of f at a point $x \in D$ is a vector $s \in \mathbb{R}^n$ satisfying the **subgradient inequality**

$$f(y) \geq f(x) + \langle s, y - x \rangle, \quad \forall y \in D.$$

The set of all subgradients of f at x is the **subdifferential** at x, denoted by $\partial f(x)$. x is a minimizer of f if and only if $0 \in \partial f(x)$. A subgradient at a point x and the function value $f(x)$ together describe a supporting hyperplane to the function at the point x. In other words, they give rise to a linear function that minorizes f and coincides with f in x. An oracle returning $f(x)$ and some subgradient $s \in \partial f(x)$ for given x is all the information needed by subgradient methods to produce a minimizing sequence for f under some mild regularity conditions.

Formulating certain semidefinite programs as eigenvalue optimization opens possibilities to exploit the structure of the underlying matrix. If $C - \mathcal{A}^T(y)$ is well structured in the sense that multiplication of this matrix with a vector is inexpensive then its extreme eigenvalue and the corresponding eigenvector can be computed efficiently by Lanczos methods, see e.g. Golubvan and Loan [39]. Eigenvectors to the maximal eigenvalue give rise to subgradients of the objective function in (E). The subgradients are used to form a simplified model of f that is easier to solve. The solutions of the model yield new candidates at which f is evaluated again. If the objective value decreases the algorithm moves on to the new point, otherwise the new subgradient is used to improve the model. This is the basic idea of the spectral bundle method. For further details we refer to Helmberg [42].

The primal problem (P) contains only equality constraints, hence coordinates of the corresponding dual vector y in (D) are unconstrained. Such problems are tackled by the standard spectral bundle method. Since most semidefinite relaxations to combinatorial problems contain inequalities in their basic formulation, we present the extension of the algorithm to the spectral bundle method with bounds. It solves semidefinite problems with bounded feasible sets. We summarize the main steps of the spectral bundle algorithm with bounds in Algorithm A.2.

Algorithm A.2 Spectral Bundle Method with Bounds

Input $y^0 \in \mathbb{R}^m_+$, $\varepsilon \geq 0$, $\overline{\kappa} \in (0, \infty]$, $\kappa \in (0,1)$, $\tilde{\kappa} \in (\kappa, 1)$, a weight $u > 0$.

1: $k := 0$; $\hat{y}^0 := y^0$; Compute $f(y^0)$ and $\widehat{\mathcal{W}}^0$; $\eta^0 := \eta^0_{\max}(W)$ for some $W \in \widehat{\mathcal{W}}^0$;

2: **trial point finding:**

3:      Find $W^+ \in \operatorname{Arg\,max}_{W \in \widehat{\mathcal{W}}^k} \psi(W, \eta^k)$;

4:      $\eta^+ := \eta^k_{\max}(W^+)$; $y^+ := y(W^+, \eta^+)$;

5:      **stopping criterion: if** $f(\hat{y}^k) - f_{W^+, \eta^+}(y^+) \leq \varepsilon(|f(\hat{y}^k)| + 1)$ **then terminate**;

6:      **if** $f_{W \in \widehat{\mathcal{W}}^k, \eta^+}(y^+) - f_{W^+, \eta^+}(y^+) > \overline{\kappa}[f(\hat{y}^k) - f_{W^+, \eta^+}(y^+)]$

       then $\eta^k := \eta^+$; **goto** Step 3;

7:      $y^{k+1} := y^+$; $W^{k+1} := W^+$; $\eta^{k+1} := \eta^+$;

8: **descent test:** Find $W^{k+1}_s \in \mathcal{W}$ such that either

     (a) $f(\hat{y}^k) - f_{W^{k+1}_s, 0}(y^{k+1}) \leq \tilde{\kappa}[f(\hat{y}^k) - f_{W^{k+1}, \eta^{k+1}}(y^{k+1})]$, or

     (b) $f(y^{k+1}) = f_{W^{k+1}_s, 0}(y^{k+1})$ and $f(\hat{y}^k) - f(y^{k+1}) \geq \kappa[f(\hat{y}^k) - f_{W^{k+1}, \eta^{k+1}}(y^{k+1})]$

     if (a) **then** $\hat{y}^{k+1} := \hat{y}^k$; (null step) **else** $\hat{y}^{k+1} := y^{k+1}$; (descent step)

9: **model update:** Choose a $\widehat{\mathcal{W}}^{k+1} \supset \{W^{k+1}, W^{k+1}_s\}$ of the form (A.7);

10: $k := k + 1$; **goto** Step 2;

Output Last trial point y^k and matrix W^k.

Consider a constrained semidefinite program in the form

$$\max \langle C, X \rangle,$$

$$(\text{CP}) \qquad \text{s.t.} \quad \mathcal{A}(X) + \eta = b,$$

$$X \in \mathcal{W}, \eta \geq 0.$$

Using Lagrange multipliers $y \in \mathbb{R}^m$ one obtains the dual to (CP)

$$\min_{y \in \mathbb{R}^m} f(y),$$

where

$$\begin{aligned}
f(y) &= \sup_{(X,\eta) \in \mathcal{W} \times \mathbb{R}^m_+} \langle C, X \rangle + \langle b - \eta - \mathcal{A}(X), y \rangle \\
&= \langle b, y \rangle + \sup_{W \in \mathcal{W}} \langle C - \mathcal{A}^T(y), W \rangle + \sup_{\eta \in \mathbb{R}^m_+} \langle -\eta, y \rangle.
\end{aligned}$$

Using (A.6) we rewrite the first supremum as the maximum eigenvalue function. The second supremum yields the indicator function

$$\imath_{R^m_+}(y) = \left\{ \begin{array}{ll} 0, & y \in R^m_+, \\ \infty, & y \notin R^m_+ \end{array} \right. .$$

Thus the effective domain of f is $R^m_+ =: Y$ and hence f takes the form

$$f(y) = \langle b, y \rangle + \lambda_{\max}(C - \mathcal{A}^T(y)) + \imath_Y(y).$$

Following the characterization (A.6) the function f is the supremum over the family of linear functions

$$f_{W,\eta}(y) := \langle C - \mathcal{A}^T(y), W \rangle + \langle b - \eta, y \rangle,$$

where $W \in \mathcal{W}$ are the subgradients of $\lambda_{\max}(\cdot)$ and $\eta \in Y$ are the Lagrange multipliers to the nonnegativity constraints of y, i.e., subgradients of $\imath_Y$. The subdifferential of f at $y \in Y$ is

$$\partial f(y) = \{\nabla f_{W,\eta} : f_{W,\eta}(y) = f(y), \quad (W,\eta) \in \mathcal{W} \times Y\}$$

and gives rise to the **cutting plane models** of f described by

$$f_{\widehat{\mathcal{W}},\eta}(y) := \sup_{(W,\eta) \in \widehat{\mathcal{W}} \times Y} f_{W,\eta}(y) \leq f(y) \quad \text{for all} \quad \widehat{\mathcal{W}} \subseteq \mathcal{W}, \, y \in \mathbb{R}^m.$$

The proximal bundle approach of Kiwiel [59] is applied within our settings as follows. At iteration k a new candidate y^{k+1} is determined as minimizer of the **augmented model**

$$f^k(y) := f_{\widehat{\mathcal{W}}^k,\eta}(y) + \tfrac{u}{2}\|y - \hat{y}^k\|^2,$$

where $\widehat{\mathcal{W}}^k \subset \mathcal{W}$ is formed from accumulated subgradient information. The quadratic term $\|y - \hat{y}^k\|^2$ keeps the new candidate close to the last successful iterate $\hat{y}^k$ (also called **stability center**). The **weight** u provides some indirect control on this distance. Next, $f_Y(y^{k+1})$ is evaluated and a subgradient is computed. If $f_Y(y^{k+1})$ delivers a sufficient decrease of the objective value then the next iterate $\hat{y}^{k+1}$ is substituted by y^{k+1} (**descent step**). Otherwise y^{k+1} is dismissed, we set $\hat{y}^{k+1} := \hat{y}^k$, but the subgradient information is used to improve the model at y^{k+1} in $\widehat{\mathcal{W}}^{k+1}$ (**null step**).

In the spectral bundle method of Helmberg and Rendl [47] the set $\widehat{\mathcal{W}}^k$ is restricted to the form

$$\widehat{\mathcal{W}}^k = \{P_k V P_k^T + \alpha \overline{W}_k : \operatorname{tr} V + \alpha = 1, \, V \in S^+_{r_k}, \, \alpha \geq 0\}, \tag{A.7}$$

where $P_k \in M_{n,r_k}$ is an orthonormal matrix and $\overline{W}_k \in \mathcal{W}$. The set of columns of the matrix P_k is called a **bundle**, the matrix $\overline{W}_k$ is referred to as the **aggregate**, and the number of columns r_k of P_k is called the **bundle size**. For this $\widehat{\mathcal{W}}^k$ we obtain the cutting plane model

$$f_{\widehat{\mathcal{W}}^k,\eta}(y) = \max\{\lambda_{\max}(P_k^T(C - \mathcal{A}^T(y))P_k), \langle C - \mathcal{A}^T(y), \overline{W}_k\rangle\} + \langle b - \eta, y\rangle. \qquad (A.8)$$

Thus for small r_k the value of the cutting plane model can be calculated efficiently. If $\overline{W}_k = \{0_{n\times n}\}$, the set $\widehat{\mathcal{W}}^k$ corresponds to a $\binom{r_k+1}{2}$-dimensional face of the semidefinite cone due to Theorem A.14. In this case $\widehat{\mathcal{W}}^k$ might be too small to contain an optimal matrix W^*. A nonzero matrix $\overline{W}_k$ allows $\widehat{\mathcal{W}}^k$ to reach the interior of S_n^+ without a significant increase of the computing cost for the next trial point.

Given a stability center $\hat{y}^k$ and a weight u, an optimal solution to the augmented model

$$\min_{y\in\mathbb{R}^m} f_{\widehat{\mathcal{W}}^k,\eta}(y) + \tfrac{u}{2}\|y - \hat{y}^k\|^2$$

yields a new candidate. Using standard saddle-point arguments from convex analysis [78] one can show that

$$\min_{y\in\mathbb{R}^m} f_{\widehat{\mathcal{W}}^k,\eta}(y) + \tfrac{u}{2}\|y - \hat{y}^k\|^2 = \max_{(W,\eta)\in\widehat{\mathcal{W}}^k\times Y}\ \min_{y\in\mathbb{R}^m} f_{W,\eta}(y) + \tfrac{u}{2}\|y - \hat{y}^k\|^2. \qquad (A.9)$$

Furthermore, solving the right-hand side yields the left-hand side minimizer. The right-hand inner minimization over y is unconstrained and can be determined explicitly for a fixed $W \in \widehat{\mathcal{W}}^k$ and $\eta \in Y$,

$$y_{\min}^k(W) = \hat{y}^k + \tfrac{1}{u}(\mathcal{A}(W) - b + \eta). \qquad (A.10)$$

Substituting that into the right hand side of (A.9) and using the definition of $f_{W,\eta}$ one obtains the dual function to the augmented model

$$\psi(W,\eta) := \langle C, W\rangle + \langle b - \eta - \mathcal{A}(W), \hat{y}^k\rangle - \tfrac{1}{u}\|b - \eta - \mathcal{A}(W)\|^2.$$

An exact pair maximizing ψ would yield the exact candidate via (A.10). For efficiency reasons a rough approximation is computed using a coordinate-wise approach of the Gauss-Seidel method [87]. In particular, an element

$$W^+ \in \operatorname{Argmax}\{\psi(W,\hat{\eta}) : W \in \widehat{\mathcal{W}}^k\}$$

is computed for a fixed $\hat{\eta}$. W^+ is a solution to the quadratic semidefinite program

$$\min \langle C, W\rangle + \langle b - \hat{\eta} - \mathcal{A}(W), \hat{y}^k\rangle - \tfrac{1}{2u}\|b - \hat{\eta} - \mathcal{A}(W)\|^2,$$

$$\text{s.t.} \quad W = P_k V P_k^T + \alpha\overline{W}_k,$$

$$\operatorname{tr}V + \alpha = 1,$$

$$V \succeq 0,\ \alpha \geq 0,$$

tackled by the interior point method. The next η^+ is determined as the maximizer for this fixed W^+,

$$\eta^+ := \eta^k_{\max}(W^+) := \operatorname{argmax}\{\,\psi(W^+, \eta) \,:\, \eta \in Y\,\} = \max\{0, -u\hat{y}^k + b - \mathcal{A}(W^+)\,\}.$$

The corresponding approximate candidate is feasible, i.e., it lies in the effective domain Y and satisfies the complementarity condition $\langle \eta^+, y^+ \rangle = 0$ for

$$y^+ := y_{W^+, \eta^+} = \max\{\hat{y}^k - \tfrac{1}{u}(b - \mathcal{A}(W^+)), 0\}.$$

Even though several repetitions of these **trial point finding** steps are allowed, it can be shown that one iteration suffices to ensure convergence. The function value $f(y^+)$ and a subgradient are established approximately by computing the maximum eigenvalue $\lambda_{\max}(C - \mathcal{A}^T(y^+))$ and a corresponding eigenvector v using iterative methods like Lanczos method. $W_s := vv^T$ gives rise to a subgradient of f. In case the lower estimate indicates a null step W_s guarantees sufficient improvement of the model for convergence (Step 8 in Algorithm A.2). In order to guarantee progress of the algorithm after a null step the new model $\widehat{\mathcal{W}}^{k+1}$ has to contain W^+ and W_s.

For more detailed implementation issues we refer to Helmberg and Kiwiel [45].

The proof of convergence of the algorithm (see Helmberg [42]) shows the following facts for $\varepsilon = 0$:

- If the algorithm stops after finitely many steps then the final y^k is an optimal solution.

- Either a descent step is triggered after a finite number of iterations or $\hat{y}^k$ is optimal.

- In case of an infinite number of descent steps the sequence of $f(\hat{y}^k)$ satisfies

$$f(\hat{y}^k) \downarrow \inf_{y \in \mathbb{R}^m} f(y).$$

Hence the spectral bundle method is a first order method.

Theorem A.16 (Helmberg 2000 [42]) *Assume* $\operatorname{Argmin} f \neq \emptyset$ *and let* $\varepsilon = 0$. *If Algorithm A.2 terminates then the final* (W^+, η^+) *is an optimal solution of (CP). If the algorithm does not terminate, there is a subsequence* $K \subseteq \mathbb{N}$ *for that all cluster points of* $(W^k, \eta^k)_{k \in K}$ *yield optimal solutions of (CP).*

Cutting Plane Algorithm within the Spectral Bundle Method

The main advantage of the spectral bundle method is the possibility to exploit structural properties of the cost and coefficient matrix, such as sparsity or low rank representations. Its memory usage can be kept at the same order of magnitude as the input data.

Furthermore, it offers efficient warm-start possibilities, i.e., an algorithmic exploitation of a so far achieved feasible or optimal solution of one problem to a slightly altered one. The main disadvantage of this method is its poor convergence. At the beginning the objective value improves quickly, but the convergence slows down dramatically as the optimum is approached. A faster progress can be achieved by improving the relaxation itself, i.e., by means of cutting planes. Since improving the initial relaxations by cutting planes relies on primal inequality constraints it is applicable as a subroutine of the spectral bundle method with bounds.

To avoid the complex notation of an explicit formulation of the combination of the spectral bundle method with a cutting plane algorithm we just outline the main idea and refer the reader to Helmberg [44] for full details.

We want to optimize over the intersection of $\mathcal{W}$ with a polyhedron $\{X : \mathcal{A}(X) \leq b\}$ that is given by a special type of separation oracle. The spectral bundle method is a pure dual approach. Thus the primal matrix W^+ is not feasible unless it is optimal. Due to Theorem A.16 all accumulation points of the subsequence of W^k that give rise to descent steps are optimal solutions of (CP). Assume that in iteration k of Algorithm A.2 W^+ solves the optimization problem

$$\max_{W \in \widehat{\mathcal{W}}^k} \psi(W, \eta^k)$$

(Step 3). We neglect the indices and denote

$$W^+ = PVP^T + \alpha\overline{W}.$$

If its deviation from feasibility $||\mathcal{A}(W^+) + \hat{\eta} - b||$ is reasonably small then W^+ may be regarded as an acceptable approximation of the primal optimal solution. Although P, V and α are easily stored explicitly, this may not always be possible for $\overline{W} \in S_n^+$ if n is large. Depending on what information is available concerning $\overline{W}$, the matrix W^+ offers different possibilities to employ separation procedures. Obviously, if all elements of the matrix $\overline{W}$ are stored then also W^+ is available in full. If $\overline{W}$ is too big to be formed explicitly but α is relatively small then PVP^T can be used for separation.

In combinatorial optimization a natural object to study in the design of relaxations is the polytope defined as the convex hull of the feasible integral solutions in the support space, i.e., in the space of variables with nonzero cost function. A similar approach can be followed in the spectral bundle method by storing, in addition to $\mathcal{A}(\overline{W})$ and $\langle C, \overline{W} \rangle$, those elements of $\overline{W}$ that are in the support of C. In this case $\overline{W}$ can be reconstructed on the support of C and all separation routines developed for the respective polytope can also be applied to this solution.

Adding or deleting cutting planes to the primal problem formulation corresponds to adding or deleting a variable in the dual problem. Unlike general bundle methods the spectral bundle method offers possibilities for a warm-start after such an update. The convergence analysis of the methods demonstrates that changes to the model during consecutive null steps are not applicable. But after a descent step the method may start with the new point from scratch without impact on the convergence. Therefore separation routines are usually called after a descent step has been carried out. By setting a new variable to zero, the current objective value remains unchanged, i.e., the algorithm continues from exactly the point where it stopped, except that some additional freedom has been added to the cost function. Dual variables corresponding to inactive inequalities with respect to the primal approximation contribute neither to the current matrix $C - \mathcal{A}^T(y)$ nor to the the the objective value $\lambda_{\max}(C - \mathcal{A}^T(y)) + b^T y$. Therefore they can be deleted without further considerations. This means elimination of the corresponding coordinates of $\mathcal{A}(\overline{W})$ in the cutting plane model.

For the circumstantial description of the cutting plane algorithm within the spectral bundle method as well as implementation details we refer to Helmberg [44].

A.5 Known NP-hard Problems

We list below known NP-hard problems, which we mentioned or used throughout this thesis.

Problem A.17 (Graph Partitioning) *Given a weighted graph* $G = (V, E)$ *with weights* $w_v \in \mathbb{Z}_+$ *for each* $v \in V$ *and* $l_e \in \mathbb{Z}_+$ *for each* $e \in E$, *and positive integers* K *and* J.
Question: *Is there a partition of* V *into disjoint sets* $V_1, V_2, \ldots, V_m$ *such that*

$$\sum_{v \in V_i} w_v \leq K \quad for \quad 1 \leq i \leq m$$

and such that if $E' \subseteq E$ *is the edge set of edges that have their end-nodes in two different sets* V_i *then*

$$\sum_{e \in E'} l_e \leq J.$$

The NP-completeness of the problem was shown by Hyafil and Rivest [50] in 1973. It remains NP-complete for fixed $K \geq 3$ even if all nodes and edge weights are equal to 1. However, it can be solved in polynomial time for $K = 2$ by algorithms finding matching in graphs.

Problem A.18 (Traveling Salesman) *Given a set* C *of* m *cities, distances* $d(c_i, c_j) \in \mathbb{Z}_+$ *for each pair of cities* $c_i, c_j \in C$ *and a positive integer* B .
Question: *Is there a tour of* C *having length* B *or less, i.e., a permutation* $[c_{\pi(1)}, \ldots, c_{\pi(m)}]$ *of* C *such that*

$$\sum_{i=1}^{m-1} d(c_{\pi(i)}, c_{\pi(i+1)}) + d(c_{\pi(m)}, c_{\pi(1)}) \leq B ?$$

The proof of the NP-completeness of the Traveling Salesman Problem is presented by Garey and Johnson in [36].

Problem A.19 (Bin Packing) *Given a finite set U of items, a size $s_u \in \mathbb{Z}_+$ for each $u \in U$, a positive integer bin capacity B and a positive integer K.*
Question: *Is there a partition of U into disjoint sets $U_1, \ldots, U_K$ such that for each $k = 1, \ldots, K$ we have $s(U_k) \leq B$?*

For the proof of the NP-Completeness of the Bin Packing Problem we refer to Garey and Johnson [36]. The **first-fit** heuristic is one of the most popular methods for finding solutions to bin packing problems, see e.g. Ahuja, Magnanti and Orlin [76]. Each item is examined and put into the first bin where it fits. If the item does not fit in any bin containinig at least one item a new bin is introduced. The algorithm runs in $\mathcal{O}(n^2)$ time. We modify this heuristic to partition several subsets of nodes into two sets with total node weight possibly not exceeding u_τ (see Section 6.2).

Knapsack Problem

A strong restriction in the formulation of the minimum graph bisection problem gives the capacity constraint on the node weight of the clusters (2.1). This constraint alone defines the Knapsack Problem, an NP-hard problem.

Problem A.20 (Knapsack) *Given a finite set U, for each $u \in U$ a size $s_u \in \mathbb{Z}_+$ and a value $v_u \in \mathbb{Z}_+$, and positive integers B and K.*
Question: *Is there a subset $U' \subseteq U$ such that $s(U') \leq B$ and $v(U') \geq K$?*

The Knapsack Problem is NP-complete, a proof was given by Karp [55]. The problem can be formulated as a linear 0-1 optimization problem as follows.

$$\max \sum_{u \in U} v_u x_u,$$

(KP)   s.t. $$\sum_{u \in U} s_u x_u \leq B,$$

$$x \in \{0, 1\}^{|U|}.$$

If there exists a solution to (KP) with an objective function value greater than or equal to K then the knapsack problem is solved and the solution set U' contains those items u with $x_u = 1$.

Remark A.21 *Dropping the integrality constraint in (KP) and extending the domain of the x-variables to $[0, a]$, $a \in \mathbb{R}_+$, one obtains a linear programming formulation for the* **continuous bounded knapsack problem**. *See also Martello [69] .*

The convex hull of feasible solutions to (KP) defines the **0-1 knapsack polytope**

$$P := \{x \in \{0, 1\}^{|U|} : s^T x \leq B\}.$$

Remark A.22 *From Remark A.9 one concludes that the separation problem over the 0-1 knapsack polytope P is NP-complete.*

The facial structure of the 0-1 knapsack polytope is well-investigated, see for instance Weismantel [85], Nemhauser and Vance [73] or Balas [8], but still not fully understood. Below we describe the **cover inequalities**, a well-known class of inequalities defining facets of the knapsack polytope.

Definition A.23 *A set $C \subseteq U$ is a* **cover** *if $s(C) > B$. A cover is* **minimal** *if $C \setminus \{j\}$ is not a cover for any $j \in C$.*

For a cover $C \subseteq U$ the inequality

$$\sum_{j \in C} x_j \leq |C| - 1 \tag{A.11}$$

is called **cover inequality**.

Proposition A.24 *For a cover $C \subseteq U$ the cover inequality (A.11) is valid for the knapsack polytope. If C is minimal then (A.11) defines a facet of P.*

For the proof we refer e.g. to Nemhauser and Wolsey [72] or Weismantel [85].

Separation for Cover Inequalities

Let $\mathcal{C}$ be the family of cover inequalities for P. We are given a non-integral point $\bar{x} \in [0, 1]^{|U|}$, and wish to know whether $\bar{x}$ satisfies all cover inequalities. Rewriting (A.11) as

$$\sum_{j \in C} (1 - x_j) \geq 1,$$

we need to answer the question: Does there exists a set $C \subseteq U$ with $\sum_{j \in C} s_j > B$ for which

$$\sum_{j \in C} (1 - \bar{x}_j) < 1 ?$$

This can be formulated as a binary linear program where the variable $z_j = 1$ if $j \in C$ and $z_j = 0$ otherwise,

$$\min \sum_{u \in U} (1 - \overline{x}_u) z_u,$$

(CKI)   s.t. $$\sum_{u \in U} s_u z_u > B,$$

$$z \in \{0, 1\}^{|U|}.$$

If the objective value of the optimal solution is greater than 1 then all cover inequalities are satisfied by $\overline{x}$. Otherwise, the set of indices of variables equal to 1 in the optimal solution is a cover. We refer the reader to e.g. Wolsey [88] for further details.

A.6   Dijkstra's and Kruskal's Algorithms

We present two known graph algorithms which we apply for solving various subproblems in separation routines and primal heuristics. A good overview of algorithms on graphs present Ahuja, Magnanti, and Orlin [76] and Cormen, Leiserson, Rivest, and Stein [21].

Kruskal's Algorithm

Algorithm A.3 Kruskal's Minimum Spanning Forest Algorithm

Input Graph $G = (V, E)$ with edge weights $w_e \in \mathbb{R}_+$, $e \in E$.

1: initialize list s with edges in E sorted by nondecreasing weights;

2: create set $T := \emptyset$ to store the edges of the forest;

3: **for** $i = 1, \dots, |E|$ **do**

4:      $e := s[i]$;

5:      **if** $T \cup \{e\}$ does not build a cycle in G **then**

6:          $e \rightarrow T$;

7:      **end if**

8: **end for**

Output Edges in T building minimum spanning forest in G.

A well-known (greedy) algorithm for finding a minimum spanning tree or forest in a graph $G = (V, E)$ was published by Joseph Kruskal in 1956. Initially, each vertex is in its own tree in the forest. The algorithm considers each edge one by one, ordered by nondecreasing weights. If an edge $e \in E$ connects two different trees then e is added to the set of edges of the forest, and these two trees are merged into a single tree. An edge e is discarded if it connects two vertices in the same tree.

Dijkstra's Algorithm

Algorithm A.4 Dijkstra's Shortest Path Tree Algorithm

Input Graph $G = (V, E)$ with edge weights $w_e \in \mathbb{R}_+$, $e \in E$, the root node $r \in V$.

1: create array $d \in \mathbb{R}_+^{|V|}$ for storing the distance of each other node in V to r;

2: $d[r] := 0$; $d[v] := w_{rv}$ if $rv \in E$ and $d[v] := \infty$ otherwise;

3: create array $p \in \mathbb{Z}_+^{|V|}$, where $p[v]$ stores the node preceding v to r;

4: $p[r] := r$ and $p[v] = r$ for all $rv \in E$;

5: create array $l \in \{0, 1\}^{|V|}$ to label nodes already examined;

6: set $l[r] := 1$ and $l[v] := 0$ for all other $v \in V$;

7: **repeat**

8:    find node u with $d[u] = \min\{d[v] : l[v] = 0\}$; $l[u] := 1$;

9:    **for** v with $l[v] = 0$ and $uv \in E$ **do**

10:      **if** $d[v] > d[u] + w_{uv}$ **then**

11:        $d[v] := d[u] + w_{uv}$; $p[v] := u$;

12:      **end if**

13:    **end for**

14: **until** $l[v] = 1$ for all $v \in V$

Output The array p.

A well-known algorithm for finding shortest path trees in a graph with nonnegative edge weights was introduced by Dijkstra in 1959. It works as follows. To each node v a pair (d_v, p_v) is assigned such that d_v stores the distance from v to the given root node r and p_v the node preceding v on the shortest path to r. At the beginning one sets

$$
\begin{aligned}
d_r &:= 0, \\
d_v &:= \begin{cases} w_{rv}, & \forall\, v \text{ adjacent to } r, \\ \infty, & \text{otherwise}, \end{cases} \\
p_r &:= r, \\
p_v &:= r \quad \forall\, v \text{ adjacent to } r.
\end{aligned}
$$

The node r is labeled. Repeatedly, one takes a not labeled node u with the minimum distance to the root and labels it. For each node v adjacent to u with $d_v > d_u + w_{uv}$ its distance to the root is updated by setting $d_v := d_u + w_{uv}$. Furthermore, for such node v the preceding node p_v is set to u, i.e., $p_v := u$. The loop terminates as soon as all nodes are labeled. The steps are summarized in Algorithm A.4.

The algorithm can be obviously applied for finding a shortest path joining two given nodes in a graph. In this case the algorithm is usually improved to the so called *Bidirectional Dijkstra's Algorithm*. Both end-nodes of the path are labeled as start nodes and alternately the paths beginning at each of the starting nodes are updated till a common node of both paths is permanently labeled. This BiDijkstra Algorithm has the same theoretical performance as the original one, but is known to work faster in practice.

A.7 Notation

argmin/argmax	(unique) minimizing/maximizing argument of a function
Argmin/Argmax	set of minimizing/maximizing arguments of a function
$\mathrm{diag}(A)$	the vector with components being the diagonal elements of matrix A
$\mathrm{Diag}(v)$	the matrix having on its main diagonal the components of the vector v
$\mathrm{vec}(A)$	the vector representation of matrix A
e	the vector with all entries equal to 1 of an appropriate dimension
e_i	the unit vector with 1 at the position i
E_V, $E(V)$	set of edges joining nodes in the set V
V_E, $V(E)$	set of end-nodes of edges in E
(V, E_V)	a graph induced by the node set V
(V_E, E)	a graph induced by the edge set E
H_V, H_E	the node and edge set of a graph H
$\mathcal{M}$	mutation type in the genetic heuristic
$\mathcal{R}$	the set of minimal roots defined on page 56
$P_{\mathcal{B}}$	the bisection cut polytope defined on page 9
$P_{\mathcal{K}}$	the knapsack polytope defined on page 13
$P_{\mathcal{E}}$	the equicut polytope defined on page 10
$P_{\mathcal{F}}$	the polytope associated with the NCGPP defined on page 13

Appendix B

Tables with Computational Results

Below we list tables with results summarized and discussed in Chapter 7.

graph	# nodes	# edges	node weight max	node weight mean	edge weight max	edge weight mean	degree max	degree mean	dens.	δ	alg. conn.
taq	170	424	48	6.22	110	20.36	20	5.0	2.95	9	0.06
taq	228	692	48	4.57	55	8.32	17	6.1	2.67	11	0.14
taq	334	3763	50	3.17	79	2.15	144	22.5	6.77	7	33.00
taq	1021	2253	50	3.95	20	2.00	14	4.4	0.43	14	0.07
taq	1021	5480	50	3.95	79	5.26	144	10.7	1.05	8	0.19
taq	1021	31641	50	3.95	10	1.32	284	62.0	6.08	7	0.58
diw	166	507	50	3.98	18	4.82	18	6.1	3.70	14	0.03
diw	681	1494	50	3.78	12	2.06	22	4.4	0.65	14	0.10
diw	681	3104	50	3.78	36	6.03	31	9.1	1.34	13	0.07
diw	681	6402	50	3.78	6	1.21	85	18.8	2.76	10	0.16
dmxa	1755	3686	48	2.76	10	2.03	20	4.2	0.24	20	0.02
dmxa	1755	10867	48	2.76	5	1.24	38	12.4	0.71	12	0.09
gap	2669	6182	9	2.53	6	1.99	9	4.6	0.17	16	0.01
gap	2669	24859	9	2.53	3	1.17	135	18.6	0.70	10	21.60
alue	6112	16896	8	4.15	2	1.08	9	5.5	0.09	21	0.01
alut	2292	6329	8	4.15	2	1.07	9	5.5	0.24	18	0.03
alut	2292	49450	8	4.15	3	1.15	979	431.5	18.83	5	81.00

δ - diameter

Table B.1: The characterization of *vlsi*-graphs.

graph	# nodes	# edges	density	degree max	degree mean	diameter	algebraic connectivity
kkt_lowt01	82	260	7.83	11	6.3	18	0.08
kkt_putt01	115	433	6.61	75	7.5	6	0.08
kkt_capt09	2063	10936	0.51	295	10.6	44	12.00
kkt_skwz02	2117	14001	0.63	1071	13.2	6	12.00
kkt_heat02	5150	19906	0.15	12	7.7	101	0.00
kkt_plnt01	2817	24999	0.63	716	17.7	12	14.00
kkt_orb11	2186	37871	1.59	1983	34.6	4	50.00
kkt_lnts02	17990	45883	0.03	13910	5.1	4	***
kkt_traj27	17148	112633	0.08	17147	13.1	2	***
kkt_traj33	20006	241947	0.12	20005	24.2	2	***

*** could not be calculated due to memory lackage

Table B.2: The characterization of the kkt-graphs.

Instance	$\mathcal{M}$	$p{=}100$ $k{=}2$ r	T_h	b_U	b_L	$p{=}200$ $k{=}2$ r	T_h	b_U	b_L	$p{=}150$ $k{=}5$ r	T_h	b_U	b_L
taq.170.424	1	9	17	55	55	6	37	55	55	5	36	55	55
	2	8	41	55	55	3	68	55	55	7	184	55	55
	3	7	30	55	55	8	133	55	55	6	117	55	55
	4	7	19	55	55	5	58	55	55	4	43	55	55
	5	8	36	55	55	8	139	55	55	6	107	55	55
taq.228.692	1	2	5	63	63	2	20	63	63	2	24	63	63
	2	5	46	63	63	2	70	63	63	2	98	63	63
	3	2	14	63	63	2	52	63	63	2	73	63	63
	4	2	9	63	63	2	33	63	63	2	49	63	63
	5	2	14	63	63	2	56	63	63	2	79	63	63
diw.681.3104	1	39	857	4537	169	23	1055	4385	159	22	1338	3278	158
	2	20	970	4078	162	18	2722	3589	146	11	2419	2628	133
	3	24	935	3545	166	18	2060	3821	161	18	2626	2697	149
	4	29	852	4251	165	20	1495	4066	151	20	1949	2970	149
	5	26	1124	3764	163	20	2411	3486	155	16	2621	2741	143
taq.1021.5480	1	16	1058	8463	196	17	1812	8245	247	15	2016	7116	232
	2	14	1574	7945	211	9	2724	7761	219	7	2947	6630	137
	3	14	1306	7237	220	11	2438	7342	188	10	2672	6570	230
	4	15	1225	7841	234	13	2119	7898	229	12	2397	6669	226
	5	16	1789	7466	227	11	2650	7125	217	8	2661	6464	144

Table B.3: Performance of the LP-based genetic algorithm depending on parameters p, k, and $\mathcal{M}$.

problem	exchange	$T(exchange)$	improve	time (sec.)	upper bound
	1-opt	-	no	0.65	366
diw681.1494	2-opt	-	no	0.68	366
	3-opt	2-opt	no	0.76	366
	4-opt	2-opt	no	0.88	366
	3-opt	-	no	56.63	244
	4-opt	-	no	6858.4	226
	3-opt	-	yes	17.25	142
	1-opt	-	no	1.44	398
taq1021.2253	2-opt	-	no	1.48	398
	3-opt	2-opt	no	1.79	398
	4-opt	2-opt	no	2.13	398
	3-opt	-	no	78.89	292
	4-opt	-	no	54476.17	268
	3-opt	-	yes	42.4	-
	1-opt	-	no	4.61	670
dmxa1755.3686	2-opt	-	no	4.84	606
	3-opt	2-opt	no	6.11	606
	4-opt	2-opt	no	7.19	606
	3-opt	-	no	572.18	462
	4-opt	-	no	***	***
	3-opt	-	yes	205.2	-
	1-opt	-	no	0.78	2015
diw681.3104	2-opt	-	no	0.89	1824
	3-opt	2-opt	no	1.21	1786
	4-opt	2-opt	no	1.45	1759
	3-opt	-	no	40.09	1301
	4-opt	-	no	8533.3	1265
	3-opt	-	yes	15.53	1061
	1-opt	-	no	0.27	1435
taq334.3763	2-opt	-	no	0.28	1435
	3-opt	2-opt	no	0.32	1426
	4-opt	2-opt	no	0.41	1424
	3-opt	-	no	6.58	1350
	4-opt	-	no	524.9	1341
	3-opt	-	no	1.03	-
	1-opt	-	no	1.01	643
diw681.6402	2-opt	-	no	1.03	643
	3-opt	2-opt	no	1.12	643
	4-opt	2-opt	no	1.24	643
	3-opt	-	no	39.16	641
	4-opt	-	no	19532.52	556
	3-opt	-	yes	21.92	360

$T(exchange)$ - The time limit set to the computation time of k-opt exchange.
*** - The heuristic did not terminate within 24 hours.

Table B.4: Performance of the GRASP-heuristic in k-opt exchange by increasing k.

problem	exchange	T(exchange)	improve	time (sec.)	upper bound
	1-opt	-	no	11.23	670
	2-opt	-	no	11.6	670
gap2669.6182	3-opt	2-opt	no	15.22	670
	4-opt	2-opt	no	20.45	670
	3-opt	-	no	2267.31	552
	4-opt	-	no	***	***
	3-opt	-	yes	743.0	-
	1-opt	-	no	8.11	400
alut2292.6329	2-opt	-	no	8.36	400
	3-opt	2-opt	no	11.15	400
	4-opt	2-opt	no	15.23	400
	3-opt	-	no	928.82	325
	4-opt	-	no	***	***
	3-opt	-	yes		-
	1-opt	-	no	1.89	3792
taq1021.5480	2-opt	-	no	1.97	3792
	3-opt	2-opt	no	2.59	3749
	4-opt	2-opt	no	3.33	3741
	3-opt	-	no	351.07	3037
	4-opt	-	no	22022.66	3037
	3-opt	-	yes	115.6	1983

Table B.5: Performance of the GRASP-heuristic in k-opt exchange by increasing k.

problem	setting	#vars	time (sec.)	b&c nodes	upper bound	h	lower bound	gap (%)
diw681.1494	**LP(IP2)**	2175	28922	239	142	g	**142.00**	0
	LP(IP3)	2152	36009	1686	144	m	140.81	2
	LP(IP3)+SDP	2152	36446	1	142	r	140.52	1
	SDP	2152	36003	237	142	r	140.49	1
taq1021.2253	LP(IP2)	3274	36132	1	118	m	114.12	3
	LP(IP3)	3259	35713	95	118	m	118.00	0
	LP(IP3)+SDP	3259	344	1	118	m	118.00	0
	SDP	3259	**322**	1	118	m	118.00	0
dmxa1755.3686	LP(IP2)	5441	36006	18	120	m	91.56	31
	LP(IP3)	5420	32879	35	**94**	g	**94.00**	**0**
	LP(IP3)+SDP	5420	36808	1	94	r	92.96	1
	SDP	5420	36014	68	94	r	92.77	1
diw681.3104	LP(IP2)	3785	36159	1	1064	m	762.48	39
	LP(IP3)	3753	36124	1	1064	m	835.66	27
	LP(IP3)+SDP	3753	36103	1	1011	r	988.79	2
	SDP	3753	36001	124	**1011**	r	**1007.07**	**0.1**

Table B.6: Performance of various models for the MGBP by solving *vlsi*-graphs.

problem	setting	#*vars*	*time* (sec.)	*b&c* *nodes*	*upper* *bound*	*h*	*lower* *bound*	*gap* (%)
taq334.3763	LP(IP2)	4097	36037	55	389	m	336.74	15
	LP(IP3)	3952	13771	351	**342**	g	**342.00**	0
	LP(IP3)+SDP	3952	27796	437	342	r	342.00	0
	SDP	3952	36001	2318	342	r	340.13	0
diw681.6402	LP(IP2)	7083	36216	1	400	m	312.43	28
	LP(IP3)	6997	36002	3	357	G	315.23	13
	LP(IP3)+SDP	6997	36637	1	331	r	323.23	2
	SDP	6997	36019	159	**331**	r	**329.21**	**0.1**
gap2669.6182	LP(IP2)	8851	36310	18	78	m	73.67	5
	LP(IP3)	8841	36025	289	78	m	73.92	5
	LP(IP3)+SDP	8841	**1495**	1	**74**	r	**74.00**	**0**
	SDP	8841	20949	27	74	r	74.00	0
alut2292.6329	LP(IP2)	8621	36410	1	77	m	70.60	9
	LP(IP3)	8611	36241	1	77	m	69.55	10
	LP(IP3)+SDP	8611	36801	1	77	m	**76.22**	1
	SDP	8611	36030	96	77	m	76.11	1
taq1021.5480	LP(IP2)	6501	36723	1	2019	m	688.70	193
	LP(IP3)	6356	36700	1	2019	m	701.17	187
	LP(IP3)+SDP	6356	36363	1	1706	r	1534.77	11
	SDP	6356	36009	84	**1650**	r	**1586.92**	**3**
dmxa1755.10867	LP(IP2)	12622	36005	1	228	m	142.50	59
	LP(IP3)	12583	36003	62	157	g	144.13	8
	LP(IP3)+SDP	12583	36388	1	150	r	143.28	4
	SDP	12583	36003	79	**150**	r	**145.93**	**2**
alue6112.16896	LP(IP2)	23008	36026	1	146	m	28.52	411
	LP(IP3)	22998	36556	1	146	m	21.51	578
	LP(IP3)+SDP	22998	37436	1	136	r	135.53	0.1
	SDP	22998	36167	11	**136**	r	**135.59**	**0**
gap2669.24859	LP(IP2)	27528	12943	1	55	g	55.00	0
	LP(IP3)	27392	2525	1	55	g	55.00	0
	LP(IP3)+SDP	27392	519	1	55	r	55.00	0
	SDP	27392	**491**	1	55	r	55.00	0
taq1021.31641	LP(IP2)	32662	36317	1	426	m	335.67	26
	LP(IP3)	32377	36009	1	426	m	375.26	13
	LP(IP3)+SDP	32377	37070	1	404	r	398.75	1.2
	SDP	32377	36134	9	**404**	r	**399.05**	**1**

Table B.7: Performance of various models for the MGBP by solving *vlsi*-graphs. .

problem	setting	#vars	time (sec.)	b&c nodes	upper bound	h	lower bound	gap (%)
kkt2063.10936	LP(IP2)	12999	5046	77	6	r	6.00	**0**
	LP(IP3)	12703	2204	363	6	r	6.00	**0**
	LP(IP3)+SDP	12703	**1028**	1	6	r	6.00	**0**
	SDP	12703	36016	192	6	r	4.97	20
kkt2117.14001	LP(IP2)	16118	36168	1	567	m	539.54	5
	LP(IP3)	15046	36024	416	567	m	**562.62**	1
	LP(IP3)+SDP	15046	36816	1	567	m	556.56	2
	SDP	15046	36002	135	567	m	562.38	**1**
kkt5150.19906	LP(IP2)	25056	36844	1	156	m	28.59	445
	LP(IP3)	25043	36030	1	156	m	18.09	762
	LP(IP3)+SDP	25043	36505	1	156	m	19.52	699
	SDP	25043	39289	2	**150**	r	**145.45**	**3**
kkt2817.24999	LP(IP2)	27816	36130	355	51	G	12.32	314
	LP(IP3)	27099	36008	1836	74	m	**13.69**	440
	LP(IP3)+SDP	27099	36004	1459	**51**	r	12.68	**302**
	SDP	27099	36005	138	49	r	5.51	789
kkt2186.37871	LP(IP2)	40057	36215	1	2090	m	1877.64	11
	LP(IP3)	38073	39827	12	2090	m	1674.14	24
	LP(IP3)+SDP	38073	75956	56	2077	r	2066.28	2
	SDP	38073	36063	9	2077	r	**2068.30**	**1**
kkt17148.112633	LP(IP2)	129781	40974	1	8258	G	8130.97	2
	LP(IP3)	112633	41107	1	8271	m	**8146.28**	2
	LP(IP3)+SDP	112633	45855	1	8194	r	8146.00	**1**
	SDP	112633	36591	2	**8194**	r	8074.68	2
kkt20006.241947	LP(IP2)	241947	37670	1	9684	m	9503.00	1.2
	LP(IP3)	241947	37656	1	9684	m	9503.00	1.2
	LP(IP3)+SDP	241947	41965	1	9611	r	9503.00	**1.1**
	SDP	241947	37003	2	9611	r	9323.13	3

Table B.8: Performance of various models for the MGBP by solving *kkt*-graphs.

problem	relax.	time (sec.)	b&c nodes	upper bd	lower bd	gap (%)
diw681.1494	LP(IP3)	36009	1686	144	140.81	2
	SDP	36006	577	144	141.99	1
taq1021.2253	LP(IP3)	35113	172	120	120.00	**0**
	SDP	36000	561	120	118.93	0.1
dmxa1755.3686	LP(IP3)	31771	278	94	94.00	0
	SDP	**518**	1	94	94.00	0
diw681.3104	LP(IP3)	36242	1	1061	868.38	22
	SDP	36026	108	1034	1018.74	**1**
taq334.3763	LP(IP3)	7417	212	347	347.00	0
	SDP	**3997**	343	347	347.00	0
diw681.6402	LP(IP3)	36197	1	419	333.90	25
	SDP	36010	128	347	338.93	**2**
gap2669.6182	LP(IP3)	16059	1	78	78.00	0
	SDP	**5295**	3	78	78.00	0
alut2292.6329	LP(IP3)	36260	1	94	71.96	30
	SDP	36054	58	82	80.42	**1**
taq1021.5480	LP(IP3)	36026	1	2120	737.59	187
	SDP	36001	77	1706	1617.43	**5**
dmxa1755.10867	LP(IP3)	29667	5	150	150.00	0
	SDP	**4222**	11	150	150.00	0
alue6112.16896	LP(IP3)	36180	1	162	22.21	629
	SDP	36032	6	154	138.13	**11**
gap2669.24859	LP(IP3)	36138	1	115	81.78	40
	SDP	36014	70	92	82.44	**11**
taq1021.31641	LP(IP3)	36016	2	428	392.57	9
	SDP	36009	10	404	396.86	**1**
kkt2063.10936	**LP(IP3)**	6321	821	7	**7.00**	**0**
	SDP	36035	29	7	4.78	46
kkt2117.14001	**LP(IP3)**	36093	441	595	**584.82**	1
	SDP	36058	24	595	583.93	1.5
kkt5150.19906	LP(IP3)	36098	1	156	18.63	737
	SDP	36355	3	150	146.80	**2**
kkt2817.24999	**LP(IP3)**	36023	1	108	14.36	**652**
	SDP	36003	168	106	6.39	1559
kkt2186.37871	LP(IP3)	36025	13	2157	1762.21	22
	SDP	36006	4	2157	2142.90	**0**
kkt17148.112633	**LP(IP3)**	42104	1	8622	8489.33	**1**
	SDP	36700	2	8513	7037.65	20
kkt20006.241947	**LP(IP3)**	39966	1	10188	9903.00	**2**
	SDP	36597	2	9992	7896.17	26

Table B.9: Linear versus semidefinite relaxation for $\tau = 0.01$.

problem	relax.	time (sec.)	b&c nodes	upper bd	lower bd	gap (%)
diw681.1494	LP(IP3)	36002	1279	142	135.09	5
	SDP	36007	576	140	138.14	**1**
taq1021.2253	LP(IP3)	36064	1	118	111.27	6
	SDP	36003	515	118	115.23	**2**
dmxa1755.3686	LP(IP3)	36009	91	94	**91.50**	2
	SDP	36032	58	94	91.34	2
diw681.3104	LP(IP3)	36241	1	1050	788.59	33
	SDP	36001	144	1011	981.36	**3**
taq334.3763	LP(IP3)	36015	153	346	311.85	10
	SDP	**32881**	2763	328	325.11	**0**
diw681.6402	LP(IP3)	36101	1	379	293.75	29
	SDP	36003	348	304	300.89	**1**
gap2669.6182	**LP(IP3)**	**16578**	1	74	74.00	**0**
	SDP	36003	24	74	71.17	3
alut2292.6329	LP(IP3)	36039	1	77	67.10	14
	SDP	36053	83	77	75.21	**2**
taq1021.5480	LP(IP3)	36322	1	2019	661.90	205
	SDP	36000	140	1575	1544.18	**2**
dmxa1755.10867	LP(IP3)	36004	24	203	137.33	47
	SDP	36004	37	150	142.55	**2**
alue6112.16896	LP(IP3)	36379	1	142	21.47	561
	SDP	36179	9	136	133.06	**2**
gap2669.24859	**LP(IP3)**	**2747**	1	55	55.00	**0**
	SDP	36001	78	55	53.76	**2**
taq1021.31641	LP(IP3)	36009	17	407	354.12	14
	SDP	36003	14	400	391.72	**2**
kkt2063.10936	**LP(IP3)**	**8860**	3032	6	6.00	**0**
	SDP	36002	216	6	2.98	101
kkt2117.14001	LP(IP3)	36018	197	553	535.59	3
	SDP	36007	54	550	538.26	**2**
kkt5150.19906	LP(IP3)	36444	1	153	16.82	809
	SDP	36970	2	150	143.37	**4**
kkt2817.24999	LP(IP3)	6475	626	6	6.00	**0**
	SDP	**2569**	1	6	6.00	**0**
kkt2186.37871	LP(IP3)	36210	24	2048	1564.42	30
	SDP	36007	6	1975	1966.65	**0**
kkt17148.112633	LP(IP3)	40134	1	7839	7717.21	1
	SDP	36757	2	7742	7717.30	**0**
kkt20006.241947	LP(IP3)	36497	1	9234	**9003.00**	2
	SDP	36476	2	9094	9001.50	**1**

Table B.10: Linear versus semidefinite relaxation for $\tau = 0.1$.

problem	relax.	time (sec.)	b&c nodes	upper bd	lower bd	gap (%)
diw681.1494	LP(IP3)	36000	1358	140	113.33	23
	SDP	36002	747	124	121.12	**2**
taq1021.2253	LP(IP3)	36018	1	118	100.30	17
	SDP	36002	291	118	106.60	**10**
dmxa1755.3686	LP(IP3)	36023	21	88	74.46	18
	SDP	36019	62	82	78.47	**4**
diw681.3104	LP(IP3)	36191	1	923	613.57	50
	SDP	36005	475	794	783.52	**1**
taq334.3763	**LP(IP3)**	**3147**	15	239	239.00	0
	SDP	5487	479	239	239.00	0
diw681.6402	LP(IP3)	36246	24	327	225.28	45
	SDP	36002	213	243	225.06	**7**
gap2669.6182	**LP(IP3)**	**27146**	17	74	74.00	**0**
	SDP	36041	20	74	65.92	12
alut2292.6329	LP(IP3)	36061	1	77	54.69	40
	SDP	36014	86	77	70.44	**9**
taq1021.5480	LP(IP3)	36448	1	1741	508.76	242
	SDP	36025	144	1483	1371.69	**8**
dmxa1755.10867	**LP(IP3)**	36638	1	113	111.18	**1**
	SDP	36003	181	113	108.61	4
alue6112.16896	LP(IP3)	36359	1	142	15.08	841
	SDP	36054	9	136	121.47	**11**
gap2669.24859	**LP(IP3)**	**6677**	1	55	55.00	**0**
	SDP	36009	66	55	50.08	9
taq1021.31641	LP(IP3)	36015	29	402	274.00	46
	SDP	36011	22	374	364.61	**2**
kkt2063.10936	**LP(IP3)**	**190**	1	3	3.00	0
	SDP	220	1	3	3.00	0
kkt2117.14001	LP(IP3)	36014	69	448	426.37	5
	SDP	36005	57	441	429.90	**2**
kkt5150.19906	LP(IP3)	36386	1	150	13.26	1031
	SDP	36718	2	150	131.46	**14**
kkt2817.24999	**LP(IP3)**	**8939**	588	6	6.00	0
	SDP	11996	15	6	6.00	0
kkt2186.37871	LP(IP3)	36149	1	1770	1422.72	24
	SDP	36028	5	1582	1560.66	**1**
kkt17148.112633	LP(IP3)	64218	2	6081	**6002.00**	1
	SDP	36670	2	6050	6001.44	**0**
kkt20006.241947	**LP(IP3)**	37514	1	7099	**7003.00**	1
	SDP	36336	2	7091	6998.67	1

Table B.11: Linear versus semidefinite relaxation for $\tau = 0.3$.

problem	relax.	time (sec.)	b&c nodes	upper bd	lower bd	gap (%)
diw681.1494	LP(IP3)	36011	1596	134	84.48	58
	SDP	36000	302	112	100.18	**11**
taq1021.2253	LP(IP3)	36072	1	108	79.23	36
	SDP	36000	211	102	89.10	**14**
dmxa1755.3686	**LP(IP3)**	**33099**	25	62	62.00	**0**
	SDP	36001	131	62	55.90	10
diw681.3104	LP(IP3)	36057	1	660	441.86	49
	SDP	36000	468	604	589.68	**2**
taq334.3763	LP(IP3)	24750	25	239	239.00	0
	SDP	**9464**	575	239	239.00	0
diw681.6402	LP(IP3)	36154	13	297	**179.14**	65
	SDP	36001	183	215	170.09	**26**
gap2669.6182	**LP(IP3)**	36197	1	74	62.94	**17**
	SDP	36022	35	74	55.27	33
alut2292.6329	LP(IP3)	36223	1	77	42.87	79
	SDP	36056	64	77	60.50	**27**
taq1021.5480	LP(IP3)	36229	1	1570	371.46	322
	SDP	36051	110	1153	1070.82	**7**
dmxa1755.10867	**LP(IP3)**	36206	1	113	102.00	**10**
	SDP	36006	140	113	90.50	24
alue6112.16896	LP(IP3)	36667	1	142	10.95	1197
	SDP	36102	9	136	101.87	**33**
gap2669.24859	**LP(IP3)**	**7003**	1	55	55.00	**0**
	SDP	36006	60	55	41.50	32
taq1021.31641	LP(IP3)	36257	132	247	**186.53**	32
	SDP	36014	139	190	186.37	**1**
kkt2063.10936	**LP(IP3)**	**252**	1	3	3.00	**0**
	SDP	318	1	3	3.00	0
kkt2117.14001	LP(IP3)	36089	1	357	317.55	12
	SDP	36033	70	329	321.20	**2**
kkt5150.19906	**LP(IP3)**	36065	1	150	10.79	**1290**
	SDP	36207	9	150	5.58	2588
kkt2817.24999	**LP(IP3)**	**1990**	330	6	6.00	**0**
	SDP	36015	98	6	4.27	40
kkt2186.37871	LP(IP3)	36251	1	1374	1012.57	35
	SDP	36466	2	1178	1126.74	**4**
kkt17148.112633	LP(IP3)	46113	6	4384	**4287.00**	2
	SDP	36486	2	4335	4286.51	**1**
kkt20006.241947	LP(IP3)	39539	4	5239	**5003.00**	4
	SDP	36875	2	5093	4968.88	**2**

Table B.12: Linear versus semidefinite relaxation for $\tau = 0.5$.

problem	setting	#vars	+ % of vars	time	lower bound
diw681.1494	LP(IP3)	2152		18005	138.20
	+sup1	5229	142	18098	123.03
	+sup2	2411	0.1	18001	137.32
taq1021.2253	LP(IP3)	3259		18058	113.16
	+sup1	5034	54	18264	104.13
	+sup2	3477	16	8048	113.01
dmxa1755.3686	LP(IP3)	5420		18075	88.50
	+sup1	7679	41	18170	79.26
	+sup2	5830	8	18004	88.45
diw681.3104	LP(IP3)	3753		18024	725.53
	+sup1	8192	118	18117	549.95
	+sup2	4030	7	18113	689.54
taq334.3763	LP(IP3)	3952		17684	*342.00
	+sup1	4646	17	18002	326.07
	+sup2	3992	1	17014	*342.00
diw681.6402	LP(IP3)	6997		18084	310.39
	+sup1	9216	31	18391	279.19
	+sup2	7113	2	18161	309.90
alut2292.6329	LP(IP3)	8611		18047	56.44
	+sup1	10361	20	18080	39.40
	+sup2	8830	2	18168	53.97
taq1021.5480	LP(IP3)	6356		18338	648.14
	+sup1	10073	58	18102	483.69
	+sup2	6705	5	18258	613.75
dmxa1755.10867	LP(IP3)	12583		18105	142.27
	+sup1	14981	19	18083	122.90
	+sup2	12820	2	18019	141.28
kkt2063.10936	LP(IP3)	12703		18004	5.83
	+sup1	13081	3	18081	5.78
	+sup2	12780	1	18004	5.86
kkt2117.14001	LP(IP3)	15046		18014	561.37
	+sup1	21107	40	18181	496.38
	+sup2	15399	2	18001	561.12
kkt2817.24999	LP(IP3)	27099		18009	11.11
	+sup1	27323	1	18003	11.03
	+sup2	27152	0.2	18008	11.04
kkt2186.37871	LP(IP3)	38073		18408	1674.12
	+sup1	42503	11	18087	1674.09
	+sup2	38468	1	18362	1672.16

* solved to optimality

Table B.13: Support extension for a sample of *vlsi*- and *kkt*-graphs.

List of Tables

Bibliography

[1] T. Achterberg. *Constraint integer programming.* PhD thesis, Berlin University of Technology, ZIB, Takustr. 7, 14195 Berlin, Germany, 2007.

[2] T. Achterberg, T. Koch, and A. Martin. Branching rules revisited. *Operations Research Letters,* 33:42–54, 2005.

[3] D. Alevras. Small min-cut polyhedra. *Mathematics of Operations Research,* 24(1):35–49, 1999.

[4] M. Armbruster. *Branch-and-cut for a semidefinite relaxation of large-scale minimum bisection problems.* PhD thesis, Chemnitz University of Technology, Reichenheinerstr. 39, 09107 Chemnitz, 2007.

[5] M. Armbruster, M. Fügenschuh, C. Helmberg, N. Jetchev, and A. Martin. Lp-based genetic algorithm for the minimum graph bisection problem. In *Operations Research Proceedings 2005, Bremen, September 7-9, 2005,* pages 315–320. Springer, 2005.

[6] M. Armbruster, M. Fügenschuh, C. Helmberg, N. Jetchev, and A. Martin. Hybrid genetic algorithm within branch-and-cut for the minimum graph bisection problem. In *Proceedings of 6th European Conference, EvoCOP 2006, Budapest, Hungary, April 10-12, 2006,* volume 3906 of *LNCS,* pages 1–12. Springer, 2006.

[7] M. Armbruster, M. Fügenschuh, C. Helmberg, and A. Martin. On the graph bisection cut polytope. *SIAM Journal on Discrete Mathematics,* 2007. to appear.

[8] E. Balas. Facets of the knapsack polytope. *Mathematical Programming,* 8:146–164, 1975.

[9] R. Banos, C. Gil, J. Ortega, and F. G. Montoya. Multilevel heuristic algorithm for graph partitioning. In *Proceedings of Applications of Evolutionary Computing, EvoWorkshops 2003,* volume 2611 of *LNCS,* pages 143–153. Springer, 2003.

[10] F. Barahona and L. Ladányi. Branch and cut based on volume algorithm: Steiner tree in graphs and max-cut. *RAIRO, Operations Research,* 40(1):53–73, 2006.

[11] F. Barahona and A. R. Mahjoub. On the cut polytope. *Mathematical Programming,* 36(2):157–173, 1986.

[12] R. Battiti and A. A. Bertossi. Greedy, prohibition, and reactive heuristics for graph partitioning. *IEEE Transactions on Computers*, 48(4):361 – 385, 1999.

[13] J. W. Berry and M. K. Goldberg. Path optimization for graph partitioning problems. *Discrete Applied Mathematics*, 90:27–50, 1999.

[14] L. Brunetta, M. Conforti, and G. Rinaldi. A branch-and-cut algorithm for the equicut problem. *Mathematical Programming*, 78(2):243–263, 1997.

[15] T. Bui, C. Heigham, C. Jones, and T. Leighton. Improving the performance of the Kernighan-Lin and simulated annealing graph bisection algorithms. In *DAC '89: Proceedings of the 26th ACM/IEEE conference on Design automation, Las Vegas, Nevada, United States*, pages 775–778, New York, NY, USA, 1989. ACM Press.

[16] T. Bui and B. R. Moon. Genetic algorithm and graph partitioning. *IEEE Transactions on Computers*, 45(7):841–855, 1996.

[17] S. Chopra and M. R. Rao. The partition problem. *Mathematical Programming*, 59:87–115, 1993.

[18] J. Clark and D. A. Holton. *Graphentheorie, Grundlagen und Anwendungen*. Spectrum Akademischer Verlag, 1994.

[19] M. Conforti, M. Rao, and A. Sassano. The equipartition polytope I: Formulations, dimension and basic facets. *Mathematical Programming*, 49(49–70), 1990.

[20] M. Conforti, M. Rao, and A. Sassano. The equipartitioning polytope II: Valid inequalities and facets. *Mathematical Programming*, 49(71–90), 1990.

[21] T. Cormen, C. Leiserson, R. Rivest, and C. Stein. *Introduction to Algorithms*. The MIT Press, 2001.

[22] M. Cross and C. Walshaw. Mesh partitioning: A multilevel balancing and refinement algorithm. *SIAM Journal on Scientific Computing*, 22:63–80, 2001.

[23] G. B. Dantzig, D. R. Fulkerson, and S. Johnson. Solution of a large scale traveling salesman problem. *Journal of the Operations Research Society of America 2*, pages 393 – 410, 1954.

[24] C. C. de Souza. *The graph equipartition problem: optimal solutions, extensions and applications*. PhD thesis, Faculté des Sciences Appliquées, Université Catholique de Louvain, Louvain-la-Neuve, Belgium, 1993.

[25] C. C. de Souza and M. Laurent. Some new classes of facets for the equicut polytope. *Discrete Applied Mathematics*, 62(1-3):167–191, 1995.

[26] M. Dell'Amico and F. Maffioli. A new tabu search approach to the 0-1 equicut problem. Osman, Ibrahim H. (ed.) et al., Meta-heuristics: theory and applications. International conference (MIC), Breckenridge, CO, USA, 22–26 July 1995. Dordrecht: Kluwer Academic Publishers. 361-377 (1996).

[27] M. Deza and M. Laurent. *Geometry of cuts and metrics*, volume 15 of *Algorithms and Combinatorics*. Springer, 1997.

[28] H. C. Elrod, T. A. Feo, and M. Laguna. A greedy randomized adaptive search procedure for the two-partitioning problem. *Operations Research*, 42(4):677–687, 1994.

[29] T. A. Feo and M. G. C. Resende. A probabilistic heuristic for a computationally difficult set covering problem. *Operations Research Letters*, 8:67–71, 1989.

[30] C. E. Ferreira. *On combinatorial optimization problems arising in computer system design*. PhD thesis, TU Berlin, ZIB, Takustr. 7, 14195 Berlin, Germany, 1994.

[31] C. E. Ferreira, A. Martin, C. C. de Souza, R. Weismantel, and L. A. Wolsey. Formulations and valid inequalities for the node capacitated graph partitioning problem. *Mathematical Programming*, 74:247–267, 1996.

[32] C. E. Ferreira, A. Martin, C. C. de Souza, R. Weismantel, and L. A. Wolsey. The node capacitated graph partitioning problem: A computational study. *Mathematical Programming*, 81(2):229–256, 1998.

[33] C. M. Fiduccia and R. M. Mattheyes. A linear time heuristic for improving network partitions. In *Proceedings of the 19th Design Automation Conference, Las Vegas*, pages 175–181, 1982.

[34] P. O. Fjällström. Algorithms for graph partitioning: A survey. *Linkoping Electronic Articles in Computer and Information Science*, 3, 1998.

[35] J. L. Garey and J. Yellen, editors. *Handbook of graph theory*. CRC Press, 2004.

[36] M. R. Garey and D. S. Johnson. *Computers and Intractability*. W.H. Freeman and Company, 1979.

[37] C. Gentile, U. Haus, M. Köppe, G. Rinaldi, and R. Weismantel. A combinatorial algorithm for stable sets in graphs. In M. Grötschel, editor, *The Sharpest Cut*, pages 51 – 74. MPS-SIAM, 2004. Festschrift in honor of M. Padberg's 60th birthday.

[38] M. Goemans and D. P. Williamson. Improved approximation algorithms for maximum cut and satisfiability problems. *Journal of the Association for Computing Machinery*, 42:1115–1145, 1995.

[39] G. H. Golub and C. F. Van Loan. *Matrix computations*. Johns Hopkins University Press, Baltimore, MD, USA, 3rd edition, 1996.

[40] M. Grötschel. Polyedertheorie, Vorlesungsskript, 1988.

[41] M. Grötschel, L. Lovász, and A. Schrijver. *Geometric algorithms and combinatorial optimization*. Springer, Berlin, 1988.

[42] C. Helmberg. *Semidefinite programming for combinatorial optimization*. Habilitationsschrift TU Berlin, ZR 00-34, ZIB, Takustr. 7, 14195 Berlin, Germany, 2000.

[43] C. Helmberg. Semidefinite programming. *European Journal of Operational Research*, 137:461–482, 2002.

[44] C. Helmberg. A cutting plane algorithm for large scale semidefinite relaxations. In M. Grötschel, editor, *The Sharpest Cut*, pages 233–256. MPS-SIAM, 2004. Festschrift in honor of M. Padberg's 60th birthday.

[45] C. Helmberg and K. C. Kiwiel. A spectral bundle method with bounds. *Mathematical Programming*, 93(2):173–194, 2002.

[46] C. Helmberg and F. Rendl. Solving quadratic (0, 1)-problems by semidefinite programs and cutting planes. *Mathematical Programming*, 82:291–315, 1998.

[47] C. Helmberg and F. Rendl. A spectral bundle method for semidefinite programming. *SIAM Journal on Optimization*, 10(3):673–696, 2000.

[48] S. Holm and M. M. Sørensen. The optimal graph partitioning problem. *OR Spektrum*, 15(1):1–8, 1993.

[49] H. H. Hoos and T. Stützle. *Stochastic local search: foundations and applications*. Morgan Kaufmann, 2004.

[50] L. Hyafil and R. L. Rivest. Graph partitioning and constructing optimal decision trees are polyniomial complete problems. Technical report, IRIA-Laboria, Rocquencourt, France, 1973.

[51] N. Jetchev. A heuristic for finding cycle inequalities for the node capacitated graph partitioning, 2005. Bachelor Thesis, Technische Universität Darmstadt.

[52] D. S. Johnson, C. R. Aragon, L. A. McGeoch, and C. Schevon. Optimization by simulated annealing: An experimental evaluation. *Operations Research*, 37:865–892, 1989.

[53] E. L. Johnson, A. Mehrotra, and G. L. Nemhauser. Min-cut clustering. *Mathematical Programming*, 62B(1):133–151, 1993.

[54] M. Jünger, A. Martin, G. Reinelt, and R. Weismantel. Quadratic 0/1 optimization and a decomposition approach for the placement of electronic circuits. *Mathematical Programming*, 63(3):257–279, 1994.

[55] R. M. Karp. Reducibility among combinatorial problems. In R. E. Miller and J. W. Thatcher, editors, *Complexity of Computer Computations*, pages 85–103. Plenum Press, New York, 1972.

[56] G. Karypis and V. Kumar. Multilevel *k*-way partitioning scheme for irregular graphs. *Journal of Parallel and Distributed Computing*, 48(1):96–129, 1998.

[57] G. Karypis and V. Kumar. A fast and high quality multilevel scheme for partitioning irregular graphs. *SIAM Journal on Scientific Computing*, 20(1):359–392, 1999.

[58] W. Kernighan and S. Lin. An efficient heuristic procedure for partitioning graphs. *Bell Systems Technical Journal*, 49(2):291–307, 1970.

[59] K. C. Kiwiel. Proximity control in bundle methods for convex nondifferentiable minimization. *Mathematical Programming*, 46:105–122, 1990.

[60] K. Kohmoto, K. Katayaman, and H. Narihisa. Performance of a genetic algorithm for the graph partitioning problem. *Mathematical and Computer Modelling*, 38(11-13):1325–1333, 2003.

[61] M. Köppe, Q. Louveaux, and R. Weismantel. Intermediate integer programming representations based on value disjunctions. *Discrete Optimization*, 2007. to appear.

[62] B. Korte and J. Vygen. *Combinatorial Optimization, Theory and Algorithms*, volume 21 of *Algorithms and Combinatorics*. Springer, 2nd edition, 2001.

[63] M. Laurent and S. Poljak. On a positive semidefinite relaxation of the cut polytope. *Linear Algebra and its Applications*, 223/224:439–461, 1995.

[64] T. Lengauer. *Combinatorial Algorithms for Integrated Circuit Layout*. John Wiley and Sons Ltd, Chichester, 1990.

[65] J. Lin. A grasp heuristic for the minimum graph bisection problem, 2007. Bachelor Thesis, Technische Universität Darmstadt.

[66] A. Lisser and F. Rendl. Graph partitioning using linear and semidefinite programming. *Mathematical Programming*, 95B(1):91–101, 2003.

[67] L. Lovasz. On the shannon capacity of a graph. *IEEE Transactions on Information Theory*, 25:1–7, 1979.

[68] H. Maini, K. Mehrotra, C. Mohan, and S. Ranka. Genetic algorithms for graph partitioning and incremental graph partitioning. In *Proceedings of the 1994 ACM/IEEE conference on Supercomputing*, pages 449–457, New York, USA, 1994. ACM Press.

[69] S. Martello and P. Toth. *Knapsack problems - algorithms and computer implementations*. John Wiley and Sons Ltd, Chichester, 1990.

[70] A. Martin. *Integer programs with block structure*. Habilitationsschrift TU Berlin, ZR 99-03, ZIB, Takustr. 7, 14195 Berlin, Germany, 1999.

[71] Z. Michalewicz. *Genetic algorithms + data structures = evolution programs*. Springer, New York, USA, 1999.

[72] G. Nemhauser and L. Wolsey. *Integer and combinatorial optimization*. John Willey & Sons, New York, 1999.

[73] G. L. Nemhauser and P. H. Vance. Lifted cover facets of the 0-1 knapsack polytope with gub constraints. *Operation Research Letters*, 16(5):255–263, 1994.

[74] J. Nocedal and S. Wright. Numerical optimization. In P. Glynn and S. M. Robinson, editors, *Springer Series in Operation Research*. Springer, 2000.

[75] J. Puchinger and G. R. Raidl. Combining metaheuristics and exact algorithms in combinatorial optimization: A survey and classification. In *Proceedings of the First International Work-Conference on the Interplay Between Natural and Artificial Computation*, volume 3562 of *Lecture Notes in Computer Science*, pages 41–53. Springer, 2005.

[76] J. B. Orlin R. K. Ahuja, T. L. Magnanti. *Network Flows*. Prentice Hall, 1993.

[77] F. Rendl, G. Rinaldi, and A. Wiegele. Branch-and-bound algorithm for max-cut based on combining semidefinite and polyhedral relaxations. *Optimization Online*, 2006. http://www.optimization-online.org/DB_HTML/2006/11/1528.html.

[78] R. T. Rockafellar. *Convex analysis*. Princeton University Press, New York, 1970.

[79] E. Rolland, H. Pirkul, and F. Glover. Tabu search for graph-partitioning. *Annals of Operations Research*, 63:209–232, 1996.

[80] K. Schloegel, G. Karypis, and V. Kumar. Graph partitioning for high-performance scientific simulations. In *Sourcebook of parallel computing*, pages 491–541. Morgan Kaufmann Publishers Inc., San Francisco, CA, USA, 2003.

[81] C. De Simone. The cut polytope and the boolean quadratic polytope. *Discrete Mathematics*, 79:71–75, 1990.

[82] C. De Simone and G. Rinaldi. A cutting plane algorithm for the max-cut problem. *Optimization Methods and Software*, 3:195–214, 1994.

[83] A. J. Soper, C. Walshaw, and M. Cross. A combined evolutionary search and multi-level optimisation approach to graph-partitioning. *Journal of Global Optimization*, 29(2):225–241, 2004.

[84] R. J. Vanderbei. *Linear programming: foundations and extensions*. Springer, 2001.

[85] R. Weismantel. On the 0/1 knapsack polytope. *Mathematical Programming*, 77(1(A)):49–68, 1997.

[86] A. Wiegele. *Nonlinear optimization techniques applied to combinatorial optimization problems*. PhD thesis, Klagenfurt University, Klagenfurt, Austria, 2006.

[87] Wikipedia. Die freie Enzyklopädie. http://de.wikipedia.org.

[88] L. A. Wolsey. *Integer programming*. John Wiley & Sons, New York, 1998.

[89] G. M. Ziegler. *Lectures on Polytopes*. Springer, New York, 1995.

Curriculum Vitae

Marzena Fügenschuh

geboren am 26. April 1975 in Lublin, Polen.

1990 – 1994	Jan-Zamoyski-Lyzeum in Lublin
1994 – 1998	Mathematikstudium mit Schwerpunkt Didaktik an der Marie Curie-Skłodowska Universität in Lublin
1998 – 1999	Studium im Diplomstudiengang Mathematik mit Nebenfach Informatik an der Carl von Ossietzky Universität in Oldenburg Graduierung als Diplom-Mathematikerin
1999 – 2000	Wissenschaftliche Hilfskraftstelle an der Universität Oldenburg Praktikum und Aushilfstätigkeit bei Ecclesia-Versicherungsdienste
2000 – 2004	Business Analyst im Bereich Informationstechnologie für Investment Banking bei der Commerzbank Frankfurt
2004 – 2007	Wissenschaftliche Mitarbeiterin in der Arbeitsgruppe Diskrete Optimierung des Fachbereichs Mathematik an der Technischen Universität Darmstadt